D0073468

ENGINEERING ACOUSTICS AND NOISE CONTROL

ENGINEERING ACOUSTICS AND NOISE CONTROL

Conrad J. Hemond, Jr.

Professor of Mechanical Engineering
University of Hartford
West Hartford, Connecticut

PRENTICE-HALL, INC. Englewood Cliffs, New Jersey

Library of Congress Cataloging in Publication Data

Hemond, Conrad J.
Engineering acoustics and noise control.

Includes index.
1. Acoustical engineering. 2. Noise control.
I. Title.
TA365.H4 620.2 82-5428
ISBN 0-13-278911-6 AACR2

Editorial/production supervision
and interior design by *Shari Ingerman*
Cover design by *Ray Lundgren*
Manufacturing buyer: *Joyce Levatino*

Printed in the United States of America

10 9 8 7 6 5 4 3 2 1

ISBN 0-13-278911-6

Prentice-Hall International, Inc., *London*
Prentice-Hall of Australia Pty. Limited, *Sydney*
Editora Prentice-Hall do Brazil, Ltda., *Rio de Janeiro*
Prentice-Hall Canada Inc., *Toronto*
Prentice-Hall of India Private Limited, *New Delhi*
Prentice-Hall of Japan, Inc., *Tokyo*
Prentice-Hall of Southeast Asia Pte. Ltd., *Singapore*
Whitehall Books Limited, Wellington, *New Zealand*

Contents

chapter 9 INSTRUMENTATION 119

chapter 10 COMMUNITY NOISE 134

chapter 11 AIR DISTRIBUTION SYSTEM NOISE 153

Preface

This textbook has been prepared for use at the undergraduate level in an engineering curriculum. It is intended as a first exposure to the complex subject of acoustics. The text material has been designed for use in a one-semester course at the senior level. It is assumed that a course in Vibrations for Mechanical or Civil Engineering or a course in Systems Dynamics for Electrical Engineering students will be a prerequisite for the course in which this text is used. Mathematical treatment is minimized in order that the student may better understand the art, the fundamental concepts, and the continuity of acoustics. More rigorous mathematical approaches are the proper subjects of the many advanced, graduate-level textbooks in this field.

The chapters are organized so that fundamental topics are covered in the first seven chapters, while applications are covered in the last five chapters.

The material for this textbook has been gathered by the author over the past 10 years. Much of the material has been developed from class notes used in the author's course in Engineering Acoustics for undergraduates. Analogies from mechanical, electrical, and civil engineering as well as geometric optics have been freely used to present the material to students from various disciplines.

Practical applications of the material have been taken from the author's consulting practice over the past 25 years. Most of this practice has been associated with architectural applications, noise control in industry, community noise, and mechanical design.

The problems associated with each chapter have been carefully screened to illustrate specific points in the chapter material. Wherever feasible, the problems are illustrative of practical situations. These real problems are typical of those used as quiz problems by the author.

Since this is a practical, first course in acoustics type of book for both the student and the professional engineer, it uses real problems encountered in daily life, and both U.S. Customary and metric units have been used as feasibility dictates.

Acknowledgments

Special recognition must be given to the many individuals whose advice and counsel contributed to the completion of this textbook. My deep appreciation is expressed to Dean Emeritus Alexander H. Zerban for his friendly cajoling and encouragement throughout the process of preparing this material. His detailed reviews and suggestions during the preparation of the manuscript were extremely valuable, although, at times, the comments were critical in a not too subtle way. I am indebted to Dean T. Skipwith Lewis for his assistance in providing a sabbatical to initiate this project and for his continued personal support during the completion of this textbook.

Encouragement and constructive criticism from faculty colleagues and from students exposed to new rough drafts of the material each year sustained me at many times when the end product seemed almost unattainable. Students were requested to be critical in a positive fashion, and they were. Many took the assignment willingly and provided very useful, clarifying, and constructive opinions.

A special word of thanks should go to the administration of the University of Hartford for providing the academic climate which allows a professor to undertake a project of this magnitude. Provisions for stenographic service and facilities to reproduce the material are essential aids to any faculty member in a project of this kind.

A final acknowledgment must be made to my family. To them, it must seem that they have been living with this project forever. A public word of thanks for their tolerance and persistent encouragement over several years must go to my wife, Eileen, to my children, Nancy, her husband Tony, Beth, and Rick.

Conrad Hemond

ENGINEERING ACOUSTICS AND NOISE CONTROL

Part I

ENGINEERING ACOUSTICS

1

Terminology

1.1 INTRODUCTION

Sound is often defined as a pressure pulse superimposed on an existing pressure. This definition establishes the need for a medium with pressure and elasticity. The medium may be solid, liquid, or gaseous. Sound cannot be propagated in a vacuum. Sound does propagate in a mathematically definable manner depending upon the elasticity and density of a medium.

The structure of the pressure pulse may be simple or complex. The repetition rate of the pulse is variable. Using these constraints all forms of the phenomena called sound can be explained. Noise is sound that is unstructured and complex as compared with music which can be shown to be structured, although it may also have a complex structure.

Sounds may be inaudible because the repetition rate of the pressure pulse may be either below or above the frequency range of hearing. Sounds may be inaudible because the energy represented by the pressure pulse may be too low, or it may be so high as to cause pain or permanent damage to the hearing mechanism.

We will see that sound and sound propagation are forms of wave motion behaving unlike most other forms of wave motion in certain applications but exactly as expected in some applications.

The word "acoustics" applies to the study of sound over the entire range of source types, propagation modes, and receiving conditions, whether or not the sounds fall into the range of hearing of normal indi-

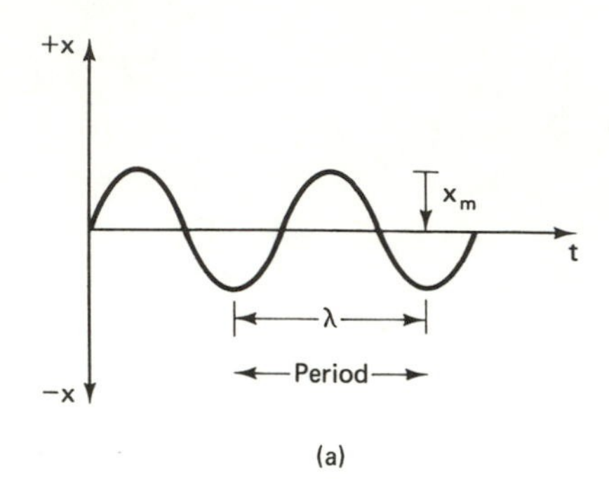

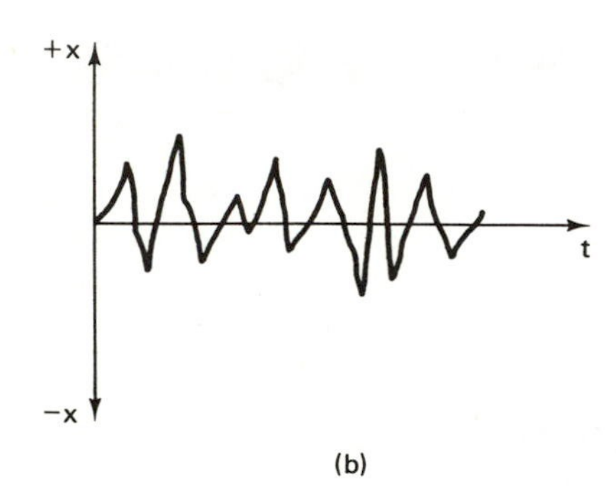

Figure 1.1 Wave forms: (a) simple, harmonic; (b) complex, aperiodic.

viduals. The word applies equally well to mechanical vibrational phenomena, subaudible and ultraaudible frequency ranges, and those phenomena associated with the sound properties of rooms and auditoriums.

By definition, sources of sound waves must be the vibration or alternate back and forth motion of some material body (molecular or macroscopic in size). The character of the sound is established by the particular motion. The motion may be simple, periodic, and of fixed excursion with time from a rest point, or complex, aperiodic, and of variable excursion distance with time from the mean or rest point. See Figure 1.1.

1.2 FREQUENCY

The rate with which an object moves back and forth can be used as a definition of its frequency; thus, the frequency is the number of vibrations occurring in a unit time. It is the number of times a complete wave form repeats itself with time. It is the number of cycles repeated in a unit of time. In the past the unit of measure was defined as cycles per second (cps). Presently, frequency is defined in units called *Hertz* (Hz). One Hertz is equivalent to one cycle per second. The accepted audible frequency range for human beings is 20 to 20,000 Hz. However, the sensitivity of the ear is not constant over that range of frequencies. This concept will be further discussed in Chapter 6. Frequencies below 20 Hz are usually referred to as subaudible or vibration frequencies (felt but not heard), and those above 20,000 Hz are referred to as ultrasonic. The reciprocal of the frequency is the time for one complete vibrational cycle. It is measured in fractions of a second, and is known as the *period*. Thus, a frequency of 20 Hz would have a period of 0.05 seconds (s).

The frequency of a sound wave indicates the number of times the compression portion of the wave passes a given point in a given time, usually 1 s. The compression portion of the wave followed by its accompanying rarefaction is caused when the pressure pulse moves through an elastic medium and causes the particles of the medium to move closer together. After the passage of the pulse, the particles of the medium attempt to seek their equilibrium position. Particle behavior is likened to a mass hung from the end of a spring. Once the mass is pulled from its position of rest and released, it tends to oscillate with a periodic or repetitive motion until the energy of the spring reaches a stable condition—usually at rest.

TABLE 1.1 Range of Frequencies

Sources	Frequency Range (Hz)
Human	85–5000
Dog	450–1080
Cat	780–1520
Piano	30–4100
Standard musical pitch (A)	440
Trumpet	190–990
Kettle drum	95–180
Bat	10,000–120,000
Cricket	7000–100,000
Robin	2000–13,000
Porpoise	7000–120,000
Jet engine	5–50,000
Automobile	15–30,000
Receivers	
Human	20–20,000
Dog	15–50,000
Cat	60–65,000
Bat	1000–120,000
Cricket	100–15,000
Robin	250–21,000
Porpoise	150–150,000

Particles in the elastic medium behave in the same manner when activated by the passage of the sound pressure pulse. The sound wave thus can be considered to create a series of compression peaks followed by a valley indicating rarefaction. The action can be visualized by watching water waves pass a stick placed in the water. The wave action causes the water to rise and fall on the surface of the stick. The rate per unit time of this rising and falling action indicates the frequency of the wave. The time for one oscillation is known as the period. Table 1.1 gives examples of the emitting and receiving frequency ranges of some common sound sources and receivers.

1.3 OCTAVE BANDS

For most engineering applications the greatest interest lies in the frequency range from 20 to 20,000 Hz. When determining the sound level of a source, a vital part of the analytical process is to know the spectrum of the sound (the distribution of sound pressure with frequency) and thus to know the particular frequencies which contribute to the highest noise levels. While it is possible to analyze a source on a frequency by frequency basis, this is both impractical and time-consuming. For this reason a scale of octave bands and one-third octave bands has been developed. Each band covers a specific range of frequencies and excludes all others. We accomplish this in the detection process by the use of band-pass filters, electronic devices which permit only certain frequencies to pass through the circuit. These filters are identified by their midfrequencies and are designed to cover consecutive ranges of frequencies, as shown in Table 1.2.

The word "octave" is borrowed from muscial nomenclature where it refers to a span of eight notes, that is, do to do. The ratio of the frequency of the highest note to the lowest note in an octave is 2:1.

The center frequency for the octave band-pass filters — and one-third octave filters — can be determined by the following relationships:

If f_n is the lower band-pass filter limiting frequency and f_{n+1} is the upper band-pass filter limiting frequency the ratio of the band limits is given by

$$\frac{f_{n+1}}{f_n} = 2^k \tag{1–1}$$

where k is 1 for full octave bands and k is 1/3 for one-third octave bands.

The midfrequency of the nth octave band is given by

$$f_m = \sqrt{f_n \cdot f_{n+1}} = 1.414 f_n \tag{1–2}$$

and for one-third octaves

$$f_m = 1.12 f_n \tag{1–3}$$

Proof of these two equations is left to the student.

By international agreement, the limits in Table 1.2 were established so that the middle band has its midfrequency at 1000 Hz as shown in the box in the table. The upper and lower limits shown in the table are rounded off for ease in calculation.

1.4 PERIOD

Quite often in dealing with wave phenomena it becomes necessary to describe the time required for the completion of one cycle of motion. By definition, this time is called the period of motion. Its relationship to frequency is that it is simply the reciprocal of frequency and it is expressed by

$$T_p = 1/f \text{ (s)} \tag{1–4}$$

1.5 WAVE MOTION

The passage of energy through a medium results in a wave-type motion which develops different kinds of waves depending upon the motion of the particles in the medium. While several forms of waves exist, the following are the most common.

A *transverse* wave characterizes the passage of most forms of energy through a medium. The passage of electrical, heat, or light energy is characterized by a wave created when the particles move perpendicular to the direction of the wave motion. This is illustrated in Figure 1.2.

Sound waves, on the other hand, are unique in that the particles oscillate back and forth in the direction of the wave — thus creating the *longitudinal* wave illustrated in Figure 1.3. This motion results in an alternate compression and rarefaction of the particles of the medium as the sound wave passes a given location.

Other forms of wave motion are classified as (a) *rotational* and (b) *torsional*. The particles of rotational waves rotate about a common center. This is typical of water waves as is readily seen in the curl of the ocean wave as it strikes the beach. The particles of torsional waves move in a

TABLE 1.2 Octave and One-Third Octave Midfrequencies and Band Limits

Lower Limit f_n	Octave Midfrequency f_m	Upper Limit f_{n+1}	Lower Limit f_n	One-Third Octave Midfrequency f_m	Upper Limit f_{n+1}
			18	20	22
			22	25	28
23	31.5	45	28	31.5	35
			35	40	45
			45	50	56
45	63	90	56	63	71
			71	80	90
			90	100	112
90	125	180	112	125	141
			141	160	178
			178	200	225
180	250	355	225	250	282
			282	315	355
			355	400	450
355	500	710	450	500	560

			560	630	710
			710	800	890
710	1,000	1,400	890	1,000	1,120
			1,120	1,250	1,400
			1,400	1,600	1,780
1,400	2,000	2,800	1,780	2,000	2,240
			2,240	2,500	2,800
			2,800	3,150	3,550
2,800	4,000	5,600	3,550	4,000	4,470
			4,470	5,000	5,600
			5,600	6,300	7,080
5,600	8,000	11,200	7,080	8,000	8,900
			8,900	10,000	11,200
			11,200	12,500	14,130
11,200	16,000	22,400	14,130	16,000	17,780
			17,780	20,000	22,400
			22,400	25,000	28,260
22,400	32,000	45,200	28,260	32,000	35,560

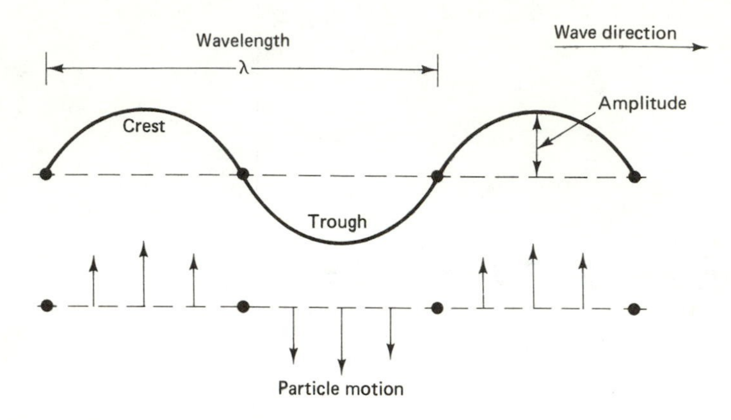

Figure 1.2 A transverse wave.

spiral form which is similar to a vector combination of longitudinal and transverse motion. These waves usually occur in solid substances and frequently result in shear patterns.

1.6 SPEED OF SOUND

The speed of a sound wave will depend upon the physical properties of the medium through which the sound wave passes. For air and most gases the speed of sound can be precisely determined by application of the general gas law of thermodynamics expressed in the following manner

$$c = \sqrt{\frac{\gamma G\, T^\circ}{M}} \tag{1–5}$$

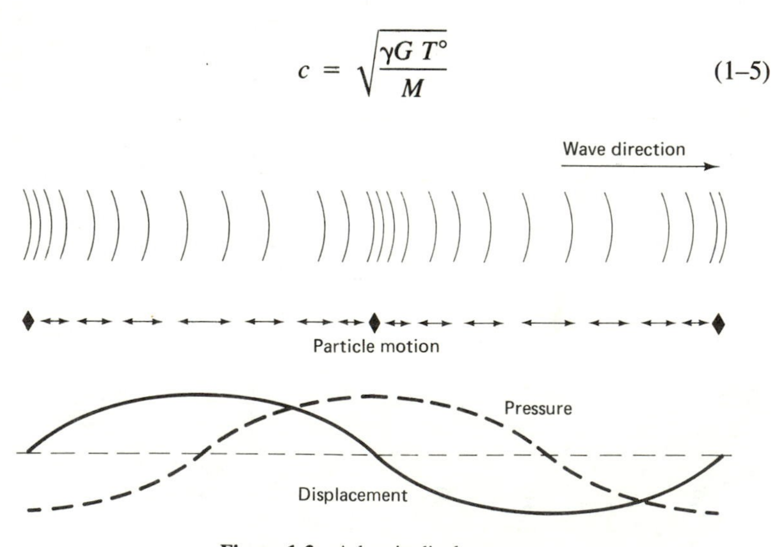

Figure 1.3 A longitudinal wave.

where c is the velocity of sound

γ is the ratio of the specific heat at constant pressure to the specific heat at constant volume

γ is 1.4 for air

G is the gas constant $= 8317\ m^2/s^2K$

$T°$ is the temperature in Kelvins (K)

and M is the molecular weight of the gas

For air at atmospheric pressure, the Equation (1–5) can be reduced to one of two forms depending upon the choice of measurement system, that is, U.S. Customary (English) or metric. For the U.S. Customary system

$$c = 49.03\sqrt{T°}\ (\text{ft/s}) \tag{1–5a}$$

where $T° =$ Rankine temperature (°R)

$= 460 +$ degrees Fahrenheit (°F)

$= 460 +$ °F

For the metric system

$$c = 20.05\sqrt{T°}\ \text{m/s} \tag{1–5b}$$

where $T° = 273 +$ degrees Celsius

$= 273 +$ °C

The speed of sound in solids is proportional to the square root of the ratio of the bulk modulus of elasticity to the density of the material. For the purposes of this text no further discussion of the speed of sound in solids is intended. We will assume the speed of sound in water is assumed to be the value for fresh water. The speed of sound in salt water is calculated from a complex empirical equation which is a cubic equation for temperature and contains corrections for other factors, such as the salinity (or salt concentration) and the depth of measurement. Students are referred to more advanced texts for specific material in this area of study.

Table 1.3 indicates the approximate speed of sound in some common materials at room temperature and normal or standard pressure.

TABLE 1.3 **Speed of Sound**

Material	ft/s	m/s
Air (at STP)	1,100	335
Lead	3,700	1128
Water (fresh)	4,500	1385
Concrete	10,200	3109
Wood (soft)	11,100	3417
Glass	15,500	4771
Steel	16,000	4925

1.7 PARTICLE MOTION

In order to achieve a proper understanding of the behavior of sound as it propagates through materials used for noise control, it is necessary to relate the motion of the particles to the parameters of the medium. The characteristic expression relating the velocity of sound c to the frequency of the wave f and to the wave length λ is expressed by the equation

$$c = f\lambda \tag{1–6}$$

In a similar manner the displacement of a particle of air due to the passage of a pressure pulse can be expressed as

$$x = \frac{u_t}{2\pi f} \tag{1–7}$$

where x is the displacement and u_t is the particle velocity.

A clear distinction should be made at this point between the velocity of propagation of a sound wave c, which is a constant in a given medium, and the particle velocity u_t which depends on the intensity of the sound source. In air, for example, the velocity of propagation at room temperature is about 340 m/s, whereas the particle velocity may vary between fractions of millimeter per second to a few centimeters per second.

The particle velocity u_t, which is the first derivative of the displacement, can be expressed as

$$u_t = 2\pi f x \tag{1–7a}$$

Also, the particle acceleration A_c, which is the first derivative of the velocity and the second derivative of the displacement, can be expressed as

$$A_c = (2\pi f)^2 x \tag{1–7b}$$

Now if we describe the resistance of a material to the passage of sound waves as the *acoustical impedance* Z (similar to electrical impedance), then the complex *specific acoustic impedance* is related to the particle velocity u_t and the dynamic pressure p_d as follows

$$Z = \frac{p_d}{u_t} \tag{1–8}$$

The units for Z are either

$$\text{Rayls} = \frac{\text{dyne-s}}{\text{cm}^3}$$

or

$$\text{MKS rayls (metric rayls)} = \frac{\text{N-s}}{\text{m}^3}$$

For a wave propagating in air in a free-field condition, the *characteristic impedance* is given by the product of the density of the medium ρ and the speed of sound c. Thus

$$Z = \frac{p_d}{u_t} = \rho c \tag{1–9}$$

The value of ρc at standard conditions of temperature and pressure is 40.7 rayls or 407 MKS rayls.

Note that, in general, where porous material surfaces are encountered by the wave, the ratio of the pressure to the particle velocity is a complex term, that is

$$Z = r_a + jx_{ar} \tag{1–10}$$

where r_a is the specific acoustic resistance and x_{ar} is the specific acoustic reactance. This subject will be discussed in a later chapter when the topic of acoustics properties of materials is presented.

Using the impedance concept, the particle motion can now be directly related to the physical properties of the material. Thus, if the particle displacement

$$x = \frac{u_t}{2\pi f} \tag{1–7}$$

then, by substitution

$$x = \frac{p_d}{2\pi f \rho c} \tag{1–11}$$

which for standard conditions, becomes

$$x = 0.0039 \frac{p_d}{f} \quad \text{(cm)} \tag{1–11a}$$

Also, the particle velocity equals

$$u_t = 2\pi f x = \frac{p_d}{\rho c} \tag{1–11b}$$

which, for ρc of 40.7 rayls,

$$u_t = 0.0246\, p_d \quad \text{(cm/s)} \tag{1–11c}$$

And the particle acceleration

$$A_c = (2\pi f)^2 x = \frac{2\pi f p_d}{\rho c} \tag{1–11d}$$

which, for ρc of 40.7 rayls

$$2\pi f A_c = 0.1545 f p_d \quad \text{(cm/s}^2\text{)} \tag{1–11e}$$

1.8 INTENSITY

The intensity of a sound wave is defined as the average rate at which sound energy is transmitted through a unit area which is positioned such that it is perpendicular to a specific direction to the point (that is, a plane perpendicular to a radius) being considered. Thus, we measure intensity at a given point located some distance from a source. See Figure 1.4. The direction of propagation from the source to the point is established. In sequence, this direction establishes the orientation of the plane of the unit area which must be perpendicular to the direction of propagation.

For most purposes the *intensity* is proportional to the square of the sound pressure and inversely proportional to the characteristic impedance, so that for a plane progressive wave

$$I = \frac{p_a^2}{\rho c} \left(\frac{\text{joules (J)}}{\text{m}^2\text{s}} \right) \tag{1–12}$$

Because sound intensities vary by large factors in the environment, it is customary to use a reference intensity based upon a reference sound pressure of 0.0002 microbar (1 microbar = 1 dyne/cm^2). Substituting this value and the previously noted value for ρc at standard conditions, the reference intensity (I_o) becomes equal to 10^{-12} W/m^2 or 10^{-16} W/cm^2. Note that the unit for pressure in the metric system is Newtons (N)/m^2 or Pascals (Pa) and 1 lb/in.^{2} equals 6894 Pa.

Also note at this time that *intensity* as a physical measurement should not be confused with *loudness* which is a subjective reaction. Loudness will be discussed in detail in Chapter 7.

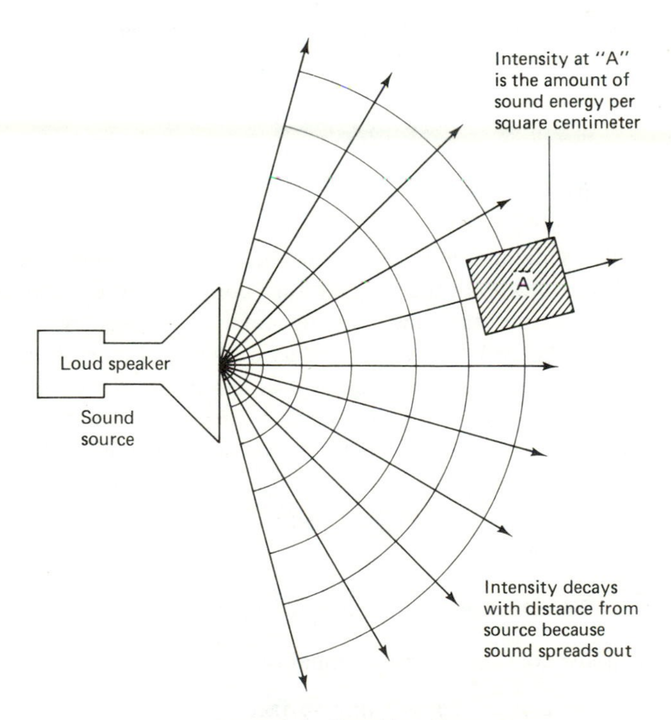

Figure 1.4 Sound intensity.

1.9 ENERGY DENSITY

A sound wave represents a form of energy which is fundamental and yet one of the most intangible quantities. It is a form of energy which dissipates into heat and is useless. It is, therefore, a wasteful and nonrecoverable form of energy. Sound energy represents the work done to displace the particles in a medium, which, in sequence, are subject to restoring forces created by the elastic medium.

Sound energy density, as the name implies, represents the energy contained in a portion of the medium divided by the volume of that portion of the medium. For plane waves, the energy density can be expressed as

$$E = \frac{I}{c} = \frac{\text{J}}{\text{m}^3} \quad \text{or} \quad \frac{\text{ergs}}{\text{m}^3} \tag{1–13}$$

Dimensional analysis shows us that I (power) is equal to W/m^2, whereas E (energy) is $\text{W/m}^2 \times [1/(\text{m/s})]$ or (J/m^3). Of course, by definition, a joule is a watt-second. Then converting joules to newton-meters so that (newton-meters/meter^3) becomes (newtons/meter^2), we end up with an expressure for *pressure*. Thus by this technique, we see that the *energy density level* of a plane progressive wave is *equal* to the *sound pressure level*.

1.10 POWER

For a simple oscillating body, the sound field can be readily determined. The sound pressure produced by vibrating body can be demonstrated to be proportional to the area of the radiating surface multiplied by the velocity of that surface. The velocity is the product of the amplitude of the surface vibration and the angular frequency of oscillation.

For a pulsating sphere whose surface area is small compared to the wavelengths of the radiated sound the following equation can be applied

$$p_{ra} = K\frac{Sv}{r} \tag{1–14}$$

where p_{ra} is the sound pressure at a radial distance r

K is the constant of proportionality

S is the area of the radiating surface

v is the velocity of the radiating surface.

It is important to note that the radiated sound pressure is a function of the velocity of the radiating surface and not the amplitude of the oscillation; the velocity, however, is the product of the amplitude and the frequency of oscillation. Hence, very small displacement amplitudes are sufficient for the radiation of high frequencies, while for efficient radiation of low frequencies, large radiating surfaces and large displacement amplitudes are required.

Now if the total sound energy radiated by a source per unit of time (second) is defined as *sound power* (usually expressed in terms of watts or ergs/s), then the sound power can be expressed as the product of the intensity at the surface multiplied by the surface area of the radiating body.

The reference sound power (W_o) can be determined by choosing a reference sphere with a small surface area S_o of 10^4 cm^2 as a standard and using the standard reference intensity I_o equal to 10^{-16} W/cm^2 so that

$$\begin{aligned} W_o &= I_o S_o \\ &= 10^{-16} \times 10^4 = 10^{-12} \end{aligned} \tag{1–15}$$

Some indication of the range of power of typical sources is indicated by the following: crickets, 5×10^{-10} W; four engines of a Boeing 707, 5×10^5 W; a Saturn rocket at launch, 5×10^7 W.

1.11 DECIBEL NOTATION

Sound energy, when compared with other forms of energy, is representative of very minute quantities in its absolute form. The range of sound energy dealt with is also much greater than the ranges dealt with in most other forms of energy. The range of typical sounds is indicated in Table 1.4.

Because the range of pressure or power in absolute units creates an almost unmanageable situation where mathematical computations are involved, the system of decibel notation has been devised. The acoustic decibel is directly comparable to the decibel as used in electrical engineering, electronics, or audio engineering, and is defined as the logarithmic ratio to the base 10 of the two quantities to be compared. Thus the decibel is defined

$$\text{dB} = 10 \log \frac{I_1}{I_o} \tag{1–16}$$

where I_1 is the intensity or power to be determined, and I_o is a reference to which all values of I are compared. Thus, if I_o represents the acoustic

TABLE 1.4 Range of Sound Pressures and Power

	lb/in.^{2}	Pascals	Watts	Horsepower
Threshold of hearing	3×10^{-9}	2.1×10^{-5}	1×10^{-12}	1.3×10^{-15}
Whisper (3 ft)	3×10^{-8}	2.1×10^{-4}	1×10^{-9}	1.3×10^{-12}
Normal speech (3 ft)	3×10^{-6}	2.1×10^{-2}	1×10^{-5}	1.3×10^{-8}
Shouting (3 ft)	3×10^{-5}	2.1×10^{-1}	1×10^{-3}	1.3×10^{-6}
Auto horn (5 ft)	3×10^{-5}	2.1×10^{-1}	1×10^{-2}	1.3×10^{-5}
Chipping hammer (3 ft)	3×10^{-4}	2.1	1.0	1.3×10^{-3}
75 piece orchestra	3×10^{-3}	21	10	1.3×10^{-2}
Medium jet engine (10 ft)	3×10^{-2}	2.1×10^{2}	1×10^{4}	13.40

power in watts at the threshold of hearing (1×10^{-12} W) the sound power level (PWL) of any other source can be represented as

$$\text{PWL(dB)} = 10 \log \frac{I_1}{1 \times 10^{-12}} \tag{1–16a}$$

As an example, the total power output of a whisper in decibels can be found

$$\text{Whisper(dB)} = 10 \log \left[\frac{1 \times 10^{-9}}{1 \times 10^{-12}} \right] = 10 \log(1 \times 10^{3})$$

$$= 30 \text{ dB}$$

Since the instrumentation used to determine sound levels consists primarily of a pressure sensitive device (microphone), it is necessary to determine the sound pressure level at the point of location of the microphone within the sound field.

Because intensity, I, is a function of pressure squared, p^2, [Equation (1–12)], we can develop an analogous equation to Equation (1–16) by substitution and write the equation for *sound pressure level* (SPL) as

$$\text{SPL} = 10 \log \frac{p_1^2}{p_o^2} \tag{1–17}$$

or

$$\text{SPL} = 20 \log \frac{p_1}{p_o} \quad \text{(dB)} \tag{1–17a}$$

where p_o, the reference pressure, is taken as the value of the sound pressure at the threshold of hearing at a frequency of 1000 Hz. For the U.S. Customary system $p_o = 3 \times 10^{-9}$ lb/in.^{2} which is equivalent to the metric standard of 0.0002 dynes/cm^2 (microbar) or 0.00002 N/m^2.

Thus if we determine that the pressure of a whisper at 3.0 ft is 3×10^{-8} lb/in.^{2}, the resulting sound pressure level will be

$$\text{SPL (dB)} = 20 \log \frac{3 \times 10^{-8}}{3 \times 10^{-9}} = 20 \log 10 = 20 \text{ dB}$$

Since PWL is a measure of the total radiated sound energy from a source, and SPL is the measure of the pressure at a radial distance x_r from the source, the relationship between the two quantities can be seen to be

$$\text{PWL} = \text{SPL} + 10 \log 2\pi x_r^2 \tag{1–18}$$

$$= \text{SPL} + 20 \log x_r + 10 \log 2\pi \tag{1–18a}$$

or, within measurement tolerance

$$\text{PWL} = \text{SPL} + 20 \log x_r + 8 \quad \text{(metric)} \tag{1–18b}$$

The above relationship is precisely valid for sources which can be considered to be radiating in a spherical pattern. For large sources with irregular radiation patterns, it is necessary to take SPL measurements at several points around the source and to average these values over a simulated hemisphere with a simulated radius of (x_{rs}) in order to correctly determine the output power. By this method, the PWL can be determined to be

$$\text{PWL} = \text{SPL}_{av} + 20 \log x_{rs} + 8 \tag{1–18c}$$

Standards for developing SPL_{av} and x_{rs} are being studied by the manufacturers and trade associations involved in the rating of large pieces of equipment which are recognized as high noise sources. Refer to trade association journals for the particulars on any given piece of equipment.

An interesting analogy can be drawn between the representation of sound intensity on the decibel scale and the concept of measuring temperature on the Celsius scale. In Figure 1.5 the somewhat arbitrary method of establishing each scale is illustrated. Arbitrarily established points on

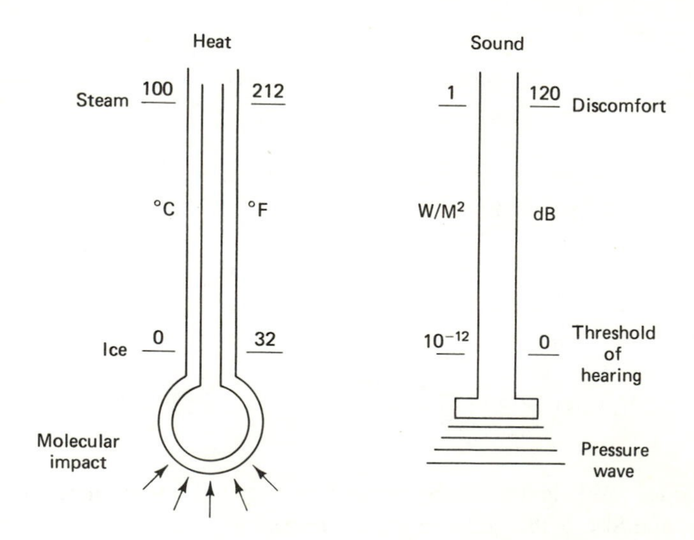

Figure 1.5 Thermometer analogs.

the Celsius scale are the melting point of ice, and the point at which steam is formed, with 100 equal divisions established between these two points. Similarly the decibel scale has an arbitrarily established zero at the threshold of hearing with 120 logarithmic divisions leading to the threshold of discomfort over an intensity range of 10^{12} W/m^2.

1.12 COMBINING DECIBELS

One of the least understood concepts concerning decibels occurs when multiple sources exist, and it becomes necessary to calculate the total pressure level at a point or to calculate the total sound power output. Since the decibel is a logarithmic function, by definition, it follows that a simple additive process cannot be applied.

While the theoretical process of combining decibels involves combining absolute pressure or intensity values, the same effect can be achieved by use of Table 1.5. This table may be used for adding any number of unlike, incoherent, or uncorrelated sound sources in decibels taken two numbers at a time. In the first column, the difference in the two levels to be combined is determined. Opposite this number in the second column is the figure for the number of decibels to be added to the *higher* level to achieve the *combined* level. Thus, for two levels with a zero difference, the combined result will increase by 3 dB. For two levels with a 3-dB difference, the combined level will be 1.75 dB higher than the highest level. Any number of decibel quantities can be added provided they are taken two at a time.

If there are two or more source levels which are identical and the

TABLE 1.5 Calculation of Combined Noise Levels

Difference Between Two Noise Levels	Number to Be Added to the Higher of the Two Noise Levels
0	3.00
1	2.54
2	2.12
3	1.75
4	1.45
5	1.18
6	0.97
7	0.78
8	0.63
9	0.51
10	0.41
20	0.05

combined level is desired then, if n is the number of sources to be combined, the sum is determined by

$$\text{dB}_{(\text{sum})} = 10 \log n + \text{dB} \tag{1–19}$$

Thus when five pieces of machinery are known to each have a fixed level of, say 85 dB, the total of the five sources will be

$$85 + 10 \log 5 = 92 \text{ dB}$$

Care must be taken in problems like this that all of the sources are equidistant from the measurement position. Refer to Appendix B for an alternate method of adding decibel levels by use of the power ratio.

PROBLEMS

1.1. The pressure level of a sound wave is 74 dB in the air.
  a. Find the absolute value of the acoustic pressure in the wave.
  b. Find the maximum value of the particle displacement, the velocity, and the acceleration of the particle at a frequency of 1000 Hz.

1.2. **a.** Compare the wavelengths of a pure tone whose frequency is 512 Hz in air, water, steel, and lead.
  b. Compare the velocity of the above tone at 60°F, 100°F, and 2000°F.

1.3. Assume that four sound sources have been measured at a distance of 10 ft and that the respective sound pressure levels are A = 95 dB, B = 92 dB, C = 89 dB, and D = 96 dB. Assume that these sources are symmetrically spaced on the circumference of a circle whose radius is 10 ft.
  a. What would be the expected sound pressure level at the center of the circle?
  b. What is the sound power level of each source?
  c. What would be the total sound power level at the center of the circle?

1.4. The sound power level of a small axial fan is 85 dB. What would be the combined sound pressure level at 30 ft if nine of these fans were in simultaneous operation?

1.5. Suppose that you wished to design a one-third octave filter about a center frequency of 440 Hz. Calculate the upper and lower limits of this band.

1.6. Standard atmospheric pressure is 14.7 lb/in^2. Calculate the pressure level in decibels of a sound creating this much over pressure. What is its intensity level?

1.7. Using Table 1.5, calculate the combined noise level of the following set of readings: 85 dB, 89 dB, 94 dB, 92 dB, 84 dB, 72 dB, and 60 dB. Calculate the same value using the method of Appendix B.

1.8. Using the following equation as the expression for the particle displacement

$$y = y_o \cos (wt - kx)$$

where $k = (\omega/c) = (2\pi/\lambda)$ (wave number), derive Equations (1–7), (1–7a), and (1–7b).

1.9. Calculate the velocity of sound in air in both the metric and U.S. Customary systems at

a. 100°F

b. 1000°F

c. 2000°F

1.10. **a.** Calculate the sound pressure level of a medium jet engine at a distance of 1000 ft assuming spherical radiation of the sound.

b. Calculate the sound power level of the same engine at the same distance. (See Table 1.4 for basic data.)

1.11. Using a loudspeaker as an example, explain why the diaphragm goes through large displacements at low frequencies and minute displacements at high frequencies for a given output power.

2

Reflection, Reverberation, and Refraction

2.1 REFLECTION

A sound wave is generally considered to propagate in a spherical pattern, moving uniformly away from a point source. An analogy is often drawn to the ripples of the surface of a pond when a stone is thrown into the water. As the radius of the spherical wave becomes very great—approaching infinity—there exists what can be referred to as a "plane" wave. Every point on a plane wave front is moving with equal speed and in the same direction. Since the behavior of any given point on the wave front is representative of the action of the wave as a whole, the path of this point can be plotted and analyzed and conclusions drawn regarding the motion of the entire wave. In a sense, borrowing from principles of geometric optics, which make great use of the tracing of rays (paths) of light, sound behavior be analyzed by use of the sound "ray." See Figure 2.1.

To help understand the principle of reflection, we will examine the simplest case. Much in the same manner as in the behavior of light, sound will be reflected from a hard surface. If we consider a plane hard surface (like that of a wall mirror), the approaching sound wave will strike the surface and a portion of its energy will be reflected in a very specific manner. If the angle between the direction of the oncoming wave and the perpendicular to the surface is defined as the incident angle ($i°$), and the angle between the reflected wave and the perpendicular to the surface as the angle of reflection ($r°$), the behavior of the sound wave can be stated

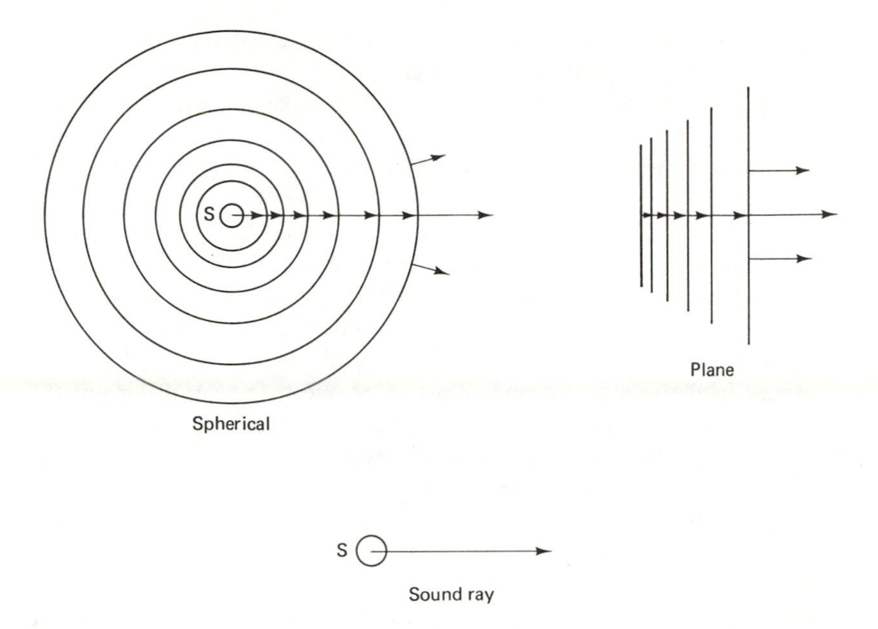

Figure 2.1 Types of sound propagation.

as follows. The angle of reflection ($r°$) will always be equal to the angle of incidence ($i°$). As indicated in Figure 2.2, both angles are defined as being measured from the normal to the surface at the point of impact.

If the surfaces are curved (convex, concave, parabolic, or compound), the principles of the action of the sound ray are identical with the plane surface (mirror) case. The only practical difficulty in some cases (particularly with compound curves) is the determination of the perpendicular (normal) to the surface. Once that is established, however, the equivalence of the reflected angle to the incidence angle can easily be determined.

The use of sound rays and the principle of reflection serves as the

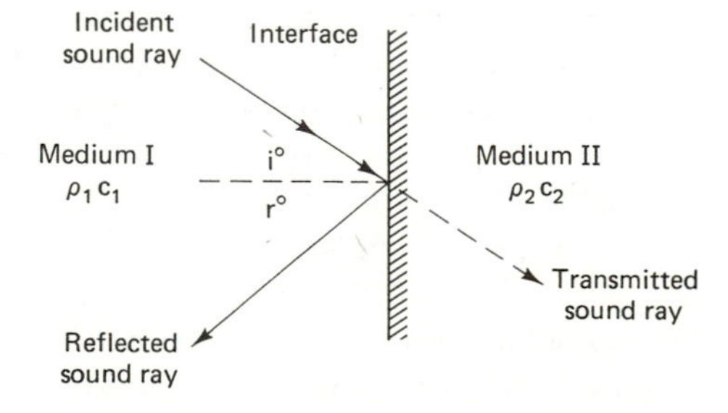

Figure 2.2 Reflection of sound off a plane surface.

basis of the graphical analysis of closed spaces, such as auditoriums. This will be further discussed later in this chapter.

To further continue the discussion of the phenomena of wave reflection, all that is fundamentally needed for the phenomena is an "interface" between two different media. In Figure 2.2, Medium I is separated from Medium II by an interface. The media have respective specific acoustic impedances of $\rho_1 c_1$ and $\rho_2 c_2$. Thus, not only do hard surfaces reflect sound but at any interface between two media with different c values, reflection phenomena will also occur. For example, sharp temperature differences are known to exist under the ocean surface at points a very short distance apart. The sharp temperature differences create sharp differences in the speed of sound. Thus ρc at one point can differ greatly from ρc at a nearby point across an existing steep thermal interface (thermocline). Signals which originate in one point of the sea will be reflected at this invisible interface as though a hard surface existed at that point. This phenomenon has caused sonar operators to get false signals from "imaginary" submarines.

Atmospheric conditions resulting from temperature inversions can also produce a sharp interface which will have the effect of bouncing an outgoing wave originating on the surface of the earth back to the ground. As will be shown later in this chapter this atmospheric phenomena is not to be confused with the principle of refraction.

It is significant at this point to note that not all of the energy of the incident wave is reflected off the interface. A portion of the energy is always transmitted across the interface. No "perfect reflectors" of sound energy are known to exist. The action of the transmitted energy will be discussed in Chapter 5.

When the behavior of sound waves in enclosed spaces is to be analyzed, the phenomena of reflection and the concept of the use of a sound ray become significant tools in determining the ideal shape for a specific acoustic purpose. The goal of such a shape study is to ensure uniform reception of sound at all points within the enclosure. Sound waves radiating from a source in an enclosed space will bounce off all surfaces. At each surface, some of the energy will be reflected and some of it will be transmitted into or through the wall surfaces (absorbed).

While noise from a source always arrives at a receiving position directly along a line-of-site path, it also arrives at that position from reflections bouncing off the ceiling and the walls. In such a *diffuse* sound field, if the reflected sound is of the same order of energy magnitude as the direct sound, listening conditions at the receiving position will be severely impaired. Information received will be garbled if the path difference is greater than 60 ft. No amount of amplification of the source will alter the situation; in fact, amplification in these cases only makes the situation worse.

It has been shown theoretically and verified experimentally that the mean free path of reflected sound waves for a large variety of room sizes and shapes is given by

$$X = \frac{4V}{S} \tag{2–1}$$

where X is the average distance that the sound travels between successive reflections, V is the volume of the space, and S is the total surface area of the room.

By use of the ray technique, a source position can be assumed on the plan of any given space and two or more (as necessary) sound rays can be traced through their sequential reflection points. In Figure 2.3 an illustration of the use of ray diagrams in the analysis of a hypothetical auditorium is shown. Potential sound problems from the nonuniform reflection of the wave are located at points A, B, and C. Direct sound is represented by solid lines while the reflections are shown as broken lines.

If the distance traveled by a sound wave along one of the reflected paths is more than 60 ft longer than the direct line of sight to a point (*A*, *B*, or *C*), there will be a distinct echo at the receiving point. The time delay or lag of approximately 0.05 s is serious enough to confuse or irritate the receiver. Such conditions lead the layperson to complain about "dead" spots or "hot" spots in poorly designed auditoriums.

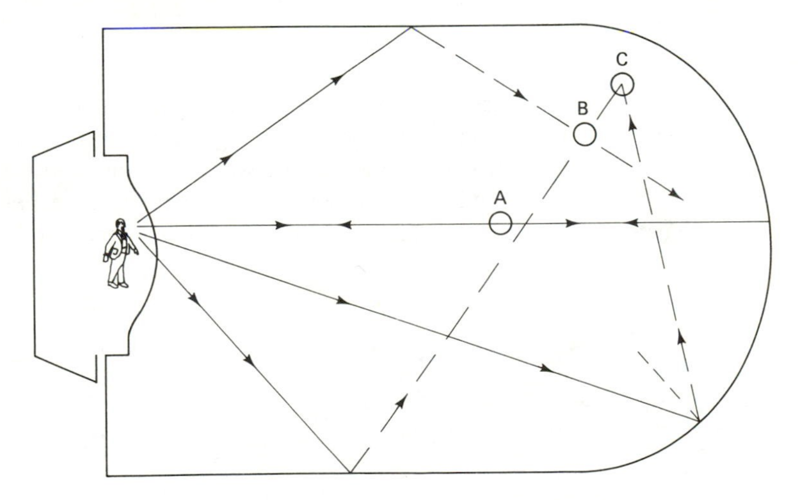

Figure 2.3 Ray diagram in a sample auditorium layout.

Referring to Figure 2.3 again, the line on which point A is located represents the possibility of a standing wave condition. Wherever there are parallel walls, an acoustic condition called standing waves can be created. An outgoing sound wave can be reflected at the wall in such a way that the returning wave is 180° out of phase with the original wave. This results in alternate reinforcement and cancellation of the original wave. Physically, a person traversing the path of the two waves will observe an alternate rise and fall of the sound pressure as he or she moves between the two walls.

Referring to Figure 2.4, if the distance, x, is an integer multiple of the wavelength, λ, a condition can arise in which the reflected wave returns exactly out of phase with the incident wave. The frequency f of the simplest mode of vibration (or the fundamental) which can occur between the two walls separated by the distance x is given by

$$f = \frac{c}{2x} = \frac{c}{\lambda} \tag{2–2}$$

Additional frequencies can arise and are represented by whole number products of Equation (2–2). Thus

$$\text{First harmonic} = f_1 = 2\left(\frac{c}{2x}\right) = \frac{c}{x} \tag{2–3}$$

$$\text{Second harmonic} = f_2 = 3\left(\frac{c}{2x}\right) = \frac{3c}{2x} \tag{2–4}$$

and

$$n\text{th harmonic} = f_n = (n + 1)\left(\frac{c}{2x}\right) = \frac{(n + 1)c}{2x} \tag{2–5}$$

Since there are a large number of wavelengths—and thus frequencies—which can satisfy the wave criteria, it is essential for good, uniform sound distribution in an enclosed space that the parallel walls be avoided in the initial design. Where construction limits dictate or the architect's desires prevail, the detrimental acoustical effects of parallelism can be minimized

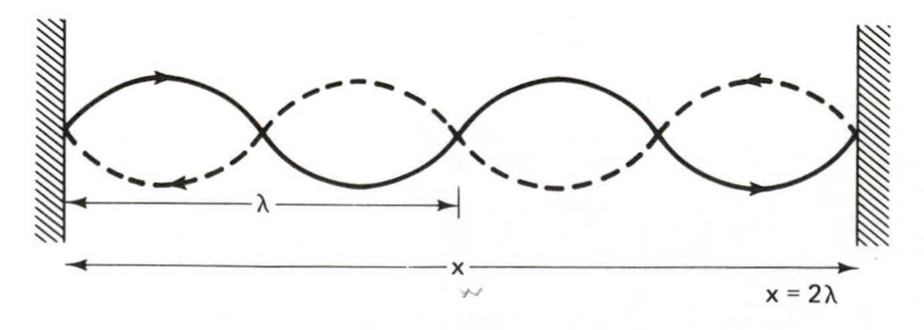

Figure 2.4 A standing wave.

by use of acoustically soft or absorbing materials on at least one of the parallel walls. Further discussion of this technique as well as techniques for eliminating other adverse sound echoes will be given in Chapter 4.

2.2 REVERBERATION

Reverberation is defined as the persistence of sound due to repeated boundary reflections after the source of sound has been stopped. It is the result of and controlled by the reflection principles discussed above. Every enclosed space exhibits this phenomena. See Figure 2.5. When the sound source is stopped, the flow of acoustic energy is still present for a finite period of time. It spreads along a number of paths with numerous reflections until the acoustic energy is exhausted. This phenomena should not to be confused with the echo effect. Sound is prolonged in spaces with hard (nonabsorbing) surfaces, that is, indoor swimming pools, large stone cathedrals, large gymnasiums, large factory buildings, and so on, long after the source of sound has ceased. Excessive reverberation, recognized by what is called the "boominess" of the space, reduces the intelligibility of speech by causing phrases and syllables to overlap, thus confusing the listener.

Since reverberation can be controlled by the use of absorbing material

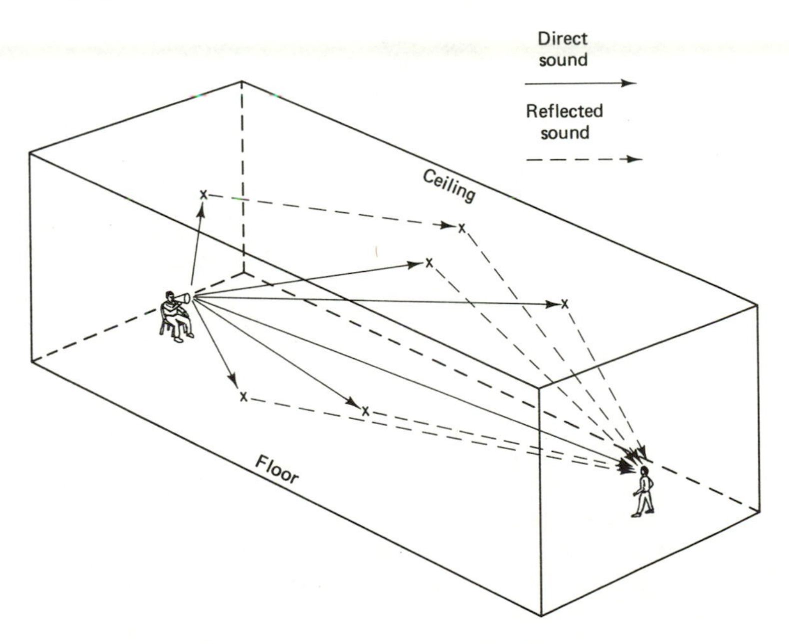

Figure 2.5 Reverberation (mixed reflections).

within a space, further discussion will be undertaken in Chapter 4. At this point it is well to indicate that the measure of reverberation is the time or rate at which the sound will decay after each reflection. This decay rate, called the *reverberation time,* is defined as the time for the original average sound pressure level to decay to a value one one-millionth of its original value. This amounts to the time for a decay of 60 dB in sound pressure. This decay rate has been shown to be a simple function of the volume of the enclosed space and the amount of acoustic absorption present. The placement or location of the absorptive material is irrelevant. Thus (in U.S. Customary units)

$$\text{Reverberation time} = T_r = \frac{0.05V}{\text{A}}\text{s} \qquad (2\text{–}6)$$

where V = volume (ft^3) and A = total absorption. Refer to Chap. 4 for a more detailed discussion.

2.3 REFRACTION

Refraction is defined as the bending of an acoustic wave due to changes in the medium through which the wave is propagating. This phenomena is not to be confused with the principle of diffraction which will be discussed in the next chapter and which occurs without any changes in the medium supporting propagation. Refraction occurs whenever those factors which determine or control the density of the medium become variable.

The bending of a sound wave passing from one medium to another is analogous to the case of light passing from air to water or air to glass. Refer back to Figure 2.2. There is a bending of the light of an angular amount which is expressed by Snell's law for the index of refraction as follows

$$\text{Index of refraction} = I = \frac{\sin i^\circ}{\sin r^\circ} \qquad (2\text{–}7)$$

where i° is the angle of incidence in the first medium (air) and r° is the angle of refraction in the second medium (water–glass), both angles being measured from the normal to the surface.

The ratio of the speed of sound in the two media will give an indication of the amount of refraction to be expected. It can be shown that the ratio of the speed of sound c_1 in Medium I to the speed of sound c_2 in Medium II is proportional to the index of refraction. Thus

$$I = \frac{\sin i^\circ}{\sin r^\circ} \propto \frac{c_1}{c_2} \tag{2–7a}$$

In enclosed spaces, refraction is rarely a phenomena to be contended with. In open spaces, however, temperature gradients usually exist near the surface of the earth. In the daytime, temperatures generally decrease with height above the ground. Since the velocity decreases with a drop in temperature, waves which leave the source with an inclination above the horizontal will bend more steeply upward as they travel out from the source. Considering a vertical cross section of the wave as it leaves the source, the upper part of the wave will be slowed down causing the entire wave to change direction and move upward. The dropoff in sound intensity with distance along the surface of the earth will hence be greater than the normal inverse square law would predict. See Figure 2.6(a).

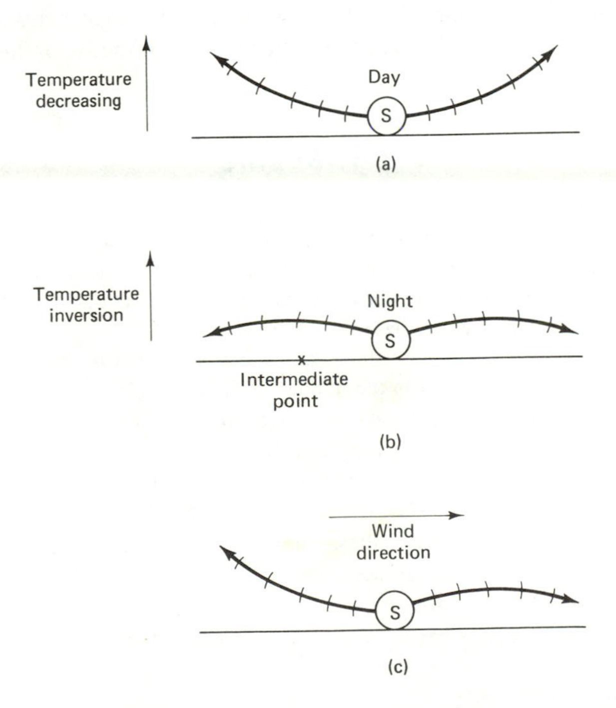

Figure 2.6 Refraction in the atmosphere.

When the temperature increases with increasing altitude (atmospheric inversion), the wave will bend downward and achieve a greater intensity at a fixed point than would normally be anticipated. This accounts for the "skip" phenomena, where under certain weather conditions, sources from great distances—10 to 15 miles—will be audible; more so than sources closer to the listener. In fact, at intermediate points the source may be completely inaudible. See Figure 2.6(b).

Wind is also a factor in the bending or refraction of sound waves. Where a sound wave moves through a wind velocity gradient, that portion of the wave passing through the high velocity portion will move ahead of the portion of the wave front lying in the slower air. Hence the wave front will tend to rotate and thus bend the path of the wave. Thus, near the Earth, erratic behavior in the propagation of sound waves is to be expected whenever the air is moving. See Figure 2.6(c). In community noise studies, for example, the wind factor as well as temperature gradients can make significant differences in the results of sound pressure measurements unless all sound readings are taken under rather constant temperature conditions and when the wind velocities are low (less than 10 miles per hour). Temperature and wind velocity should always be recorded so that any anomalies in the data can be examined in the light of these refractive influences.

PROBLEMS

2.1. Construct a rectangle 6 in. by 9 in. on a sheet of plain paper.

- **a.** Assuming a sound source at one of the corners of the rectangle, draw at least three sound rays showing at least three reflections of each. Explain what happens when these rays intersect.
- **b.** Repeat the above procedure with a source located in the middle of the short side, and then in the middle of the long side.

2.2. Assuming that your rectangle represents the floor plan of a room 60 ft by 90 ft, which has a ceiling height of 10 ft, calculate the mean free path of the sound waves within the room. Also indicate at least one standing wave condition in this room (specify the frequency and wave length).

2.3. By use of a scale drawing, illustrate how sound is reflected off

- **a.** a concave spherical surface
- **b.** a concave parabolic surface.

2.4. Using the tables of Chapter 1, calculate the index of refraction existing when sound passes

- **a.** from air to water, and
- **b.** from steel to air. If the incident angle in each case is 10°, what will be the respective angles of refraction?

2.5. Assuming that your classroom is 30 ft × 20 ft × 10 ft high and the total absorption within the room is 2500 units

a. calculate the reverberation time

b. how much absorption would be required to cut the above reverberation time in half

c. to double the reverberation time.

2.6. For the classroom in Problem 2.5 calculate the fundamental and first three harmonics of standing waves that could be supported in each of the three directions.

2.7. Explain the conditions necessary for a sound wave originating in the fire box of a furnace to pass out of an open door with a considerable reduction in energy. Is this a reasonable phenomena to be expected in practice—that is, suppose the temperature in the furnace is 3000°F and the outside temperature is 90°F.

3

Diffraction and Interference

3.1 BARRIERS

Diffraction of sound waves refers to the bending of the wave around the edges of obstructions in the path of the wave. It also refers to the creation of sound shadows behind obstacles placed in the path of the sound wave. Much of the theory of diffraction comes from the study of optics and the behavior of light. Theoretical computations of the optical effects of diffraction have been made by Rayleigh, Schwartz, Stenzel, Weiner, and others. The mathematics are complicated, requiring the application of the Fresnel integral and the Spiral of Cornu. For engineering applications involving barriers, partial height partitions, and simple obstructions, the mathematics can be reduced to simplified relationships as we will indicate.

The basic relationships of diffraction are relatively simple but not well understood by the average engineer or layperson, although the effects can be recognized by a relatively untrained observer who has some simple facts at hand.

The bending due to diffraction around a barrier is highly selective with respect to frequency. Long wavelength, low frequency sounds are less affected by the barrier than short wavelength, high frequency sounds. Thus, for a complex sound wave consisting of all frequencies, such as that associated with music, there will be uneven dispersion behind the barrier. For a listener sitting within the shadow, the bass portion of the music will predominate over the treble. This effect is most noticeable when music is played in an adjoining room with a partially intervening partition. It is this effect which also explains the failure of highway barriers or *berms* to reduce the predominantly low frequency truck noise while high frequency tire squeal may be slightly reduced in level.

In new office or school arrangements that use an *open space plan* where full height walls are eliminated and only partial height sight barriers are used, the diffraction effect negates the effectiveness of the barriers as interrupters of sound. The result of using sight in controlling the noise from one space to another is more psychological—"out of sight, out of hearing"—than physical. The sounds of low frequency voices, typewriters, and office machines will carry over the barriers to adjacent spaces. Privacy cannot be achieved by partial height partitions no matter how acoustically hard they may be.

Most of the calculations required for application of diffraction refer to the general equations for the bending of a wave around barriers, and are derived using Fresnel integrals. The equations are tabulated in many mathematical or optics texts and can be simplified as follows

$$\text{Sound level reduction (SLR)} = -3 + 10 \log [(0.5 - x_F)^2 + (0.5 - y_F)^2] \quad (3\text{–}1)$$

where x_F and y_F, the Fresnel integrals, are functions of the variable

$$\mu_F = h \sqrt{\frac{2}{\lambda}\left(\frac{1}{x_a} + \frac{1}{x_b}\right)} \quad (3\text{–}2)$$

in which h equals the height of the barrier *above the line of sight* from the source to the receiver, and x_a equals the distance from the source to the barrier x_b equals the distance from the barrier to the receiver. μ_F, x_F, and y_F are thus established by using a table of Fresnel integrals. (See Appendix C.)

Further simplification of the determination of the sound reduction properties of barriers is found in the following discussion. A barrier situation is assumed as in Figure 3.1. Here h represents the height above the *line of sight* from the source S to the potential receiver R; then the

$$\text{Sound level reduction (modified Fresnel theory) (SLR)} = -14.5 + 10 \log f(\Delta x_{AB}) \quad (3\text{–}3)$$

where $\Delta x_{AB} = x_A + x_B - x_{ab}$ and f is the desired frequency. (3–4)

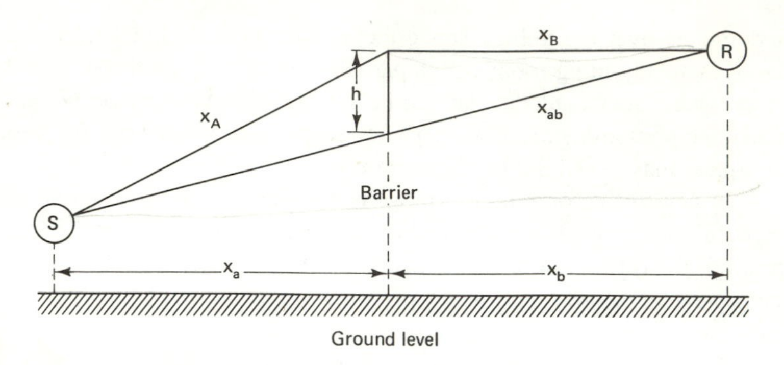

Figure 3.1 Diffraction over a barrier.

If the source-receiver line of sight is approximately horizontal, and x_a equals the distance from the source to the barrier, and x_b equals the distance from the barrier to the receiver, then the value of x_{AB} becomes equal to

$$\Delta x_{AB} = \sqrt{x_a^2 + h^2} + \sqrt{x_b^2 + h^2} - (x_a + x_b) \qquad (3\text{–}5)$$

Note here that some textbooks provide an equation for the effect of barriers for the condition "when the receiver is at a much greater distance from the wall than is the location of the source." Thus, according to those authors, if the source is at a distance R from the wall of height H, their approximate equation is

$$\text{Sound level reduction} = 10 \log \frac{10\,H^2}{\lambda R} \qquad (3\text{–}6)$$

where λ is the wavelength. This equation, which also emphasizes the inverse wavelength relationship, is only useful as an estimate since the distance from the wall to the receiver is not considered and this distance is vital to the desired end result. In fact, this equation is an abbreviated distortion of the original work of Fehr who expressed the relationship in the following manner. Fehr proposed that if R equaled the distance from the source to a barrier whose height above the line of sight was H, located at a distance D from the receiver, then the *excess* attenuation (sound level reduction) provided by the barrier could be expressed as

$$\text{Attenuation (dB)} = 10 \log (10\,N) \qquad (3\text{–}7)$$

where

$$N = \frac{2}{\lambda} R \; - (\sqrt{1 + H^2/R^2} - 1) + D (\sqrt{1 + H^2/D^2} - 1) \qquad (3\text{–}8)$$

This equation gives slightly more conservative results than the Fresnel relationship but it is often equally useful for obtaining engineering results.

From the Fresnel theory, it can be shown that the practical limit of barrier attenuation is 24 dB. This results when the receiver is at a distance of approximately 250 to 300 ft from a wall. It is at this point where the inverse square dropoff of sound with distance without the barrier negates the effect of the excess attenuation of the barrier. In other words, the path of sound over the wall is approximately equal to the source-receiver distance without the wall.

When the specific distance from the source of the potential receiver is known, the problem becomes one of locating a barrier in between the two at some appropriate distance from the source, usually fixed by the physical parameters of the site, and determining an approximate height H above the line of sight to achieve a given sound reduction at the receiver. Many times the solution technique becomes one of trial and error.

Effective use of the diffraction principles in the design of barriers requires the placement of the barrier as close to the source as possible and raising the height *above the line of sight* as much as is physically and economically feasible. It also involves using a material for the barrier whose transmission loss in decibels is at least equivalent to the anticipated wall attenuation due to diffraction. It also involves using long barriers. Sound will diffract around the ends of the barriers according to the same principles present in passage over the wall. Some authors suggest, as a rule of thumb, that the barriers must be at least four times longer than the distance from the point source to the barrier.

For barriers or partial height partitions located within structures, that is, open space offices or schools, it is vital that the relationship between the source and the receiver enter into the calculations and that the rules concerning the relationship of h to the *line of sight* be carefully observed. Further discussion will be presented in later chapters dealing with specific architectural problems.

3.2 OPENINGS

Another diffraction effect which merits discussion is the effect of cracks or small openings in a wall structure. In optics, light coming through small cracks is confined to a narrow beam of about the same size and shape as the opening. Therefore, there is very little bending of the light by the diffraction process. A crack in a water vessel on the other hand allows the water to flow out in all directions because the pressure at any point in the liquid is exerted equally in all directions at once. While not directly analogous to the water example, sound does tend to flow out of a crack in all directions.

In sound, however, the wavelengths are not small compared with the dimensions of the openings, thus the effects of diffraction will vary with frequency (that is, wavelength). In general, high frequency sound with short wavelengths will approach the optic conditions (very little bending), whereas the low frequency or long wavelength sounds will exhibit much greater bending. If the opening is large compared to the wavelength, then the diffraction effects are at a minimum. Consider, for example, a large stage opening or proscenium arch of an auditorium. Here practically no diffraction occurs. Then consider a 3-ft door opening, which is a small opening compared to the wavelength of a 100-Hz tone (whose approximate wavelength is equal to one foot). In this case diffraction will occur. The latter opening is a large opening when compared with the wavelength of

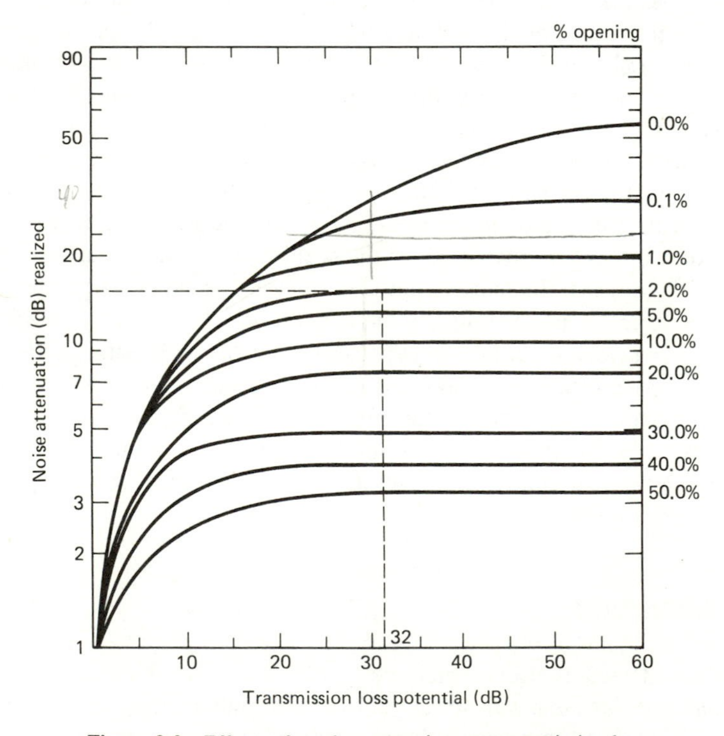

Figure 3.2 Effects of cracks or openings on transmission loss.

a 10,000-Hz tone (whose wavelength is about 1 in.) and very little bending of this tone will result.

The same diffraction effects can account for the scattering of sound by architectural projections or discontinuities in a wall. The amount of scattering resulting from a small protrusion from a flat wall will be a function of the relationship of the size of the projection to the wavelength of the sound impinging on it.

Large obstacles in the path of a sound wave will create shadows, while small obstacles will create few shadows except at very high frequencies. This factor is taken into account, for example, in the shape and design of microphones.

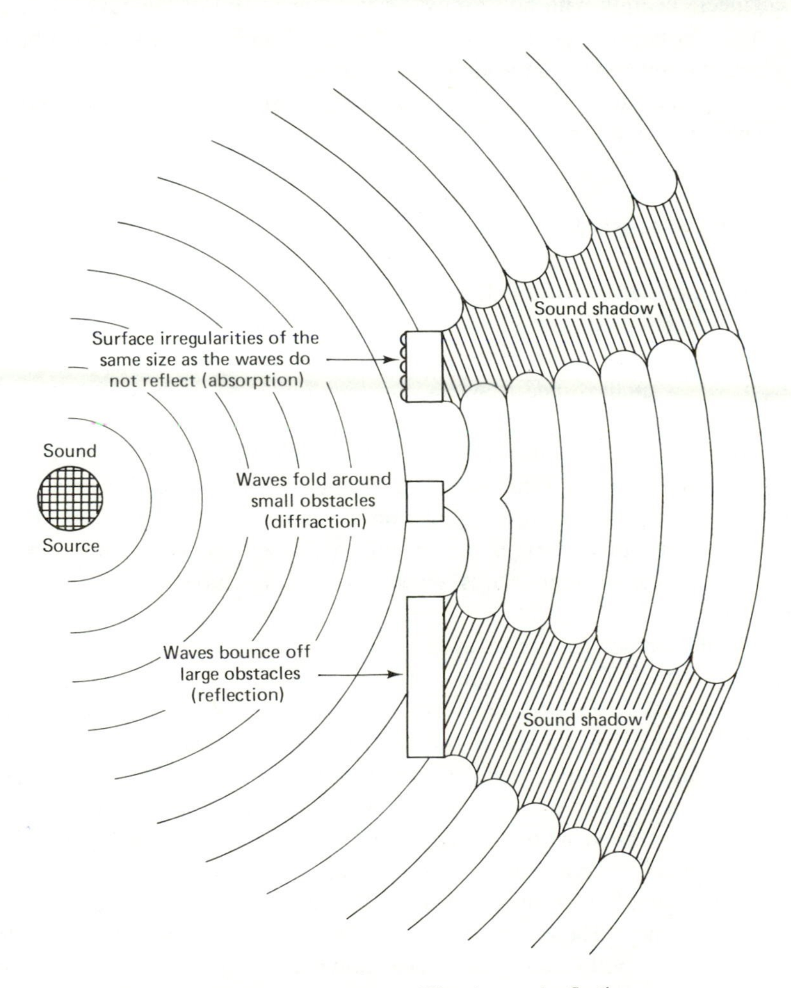

Figure 3.3 Absorption, diffraction, and reflection.

Openings cut into doors for ventilation purposes provide excellent paths for the propagation of sound. Large cuts at the bottom of doors—over the sill—for the same purpose provide excellent bypass channels for sound propagation.

Figure 3.2 indicates that if the percentage of open area in a wall is known and if the theoretical (or actual) transmission loss is known, the reduced transmission loss can be determined. This reduced loss is given by the *noise attenuation realized*. Thus if the theoretical transmission loss is 32 dB and there is a crack in the wall amounting to 2 percent, then the noise attenuation realized will be 15 dB or a total loss of 17 dB.

Figure 3.3 shows a summary of the wave phenomena with which engineers dealing with noise must be familiar. In all of the cases shown, sound originates from one source and travels directly to the specific object where its behavior is dependent on the rules which have been discussed to this point. None of these cases is to be confused with the phenomena of interference taken up in the next section.

3.3 INTERFERENCE

Interference occurs for sound waves, as in all wave motion, whenever two or more motions are simultaneously imposed upon a particle or series of particles in a medium. It refers primarily to combination effects associated with waves of the *same* frequency originating from *different* sources or from different areas of the same source.

Interference is a phenomenon that occurs in connection with almost every area or topic in acoustics. From a practical viewpoint, it is an almost useless concept for positive application to problems of noise control. It is, however, a useful concept in music and usually it is an unrecognized cause for discomfort in auditoriums where the phenomenon can occur. (See Chapter 2.)

Interference between waves (single frequency) can create constructive or destructive effects. Consider two waves of the same frequency. See Figure 3.4. Waves A and B as well as A' and B' have the same frequency, but in Figure 3.4(a) wave B is 180° out of phase with wave A. If these two waves impinge upon a particle at the same time, the net result will be the motion shown in curve C. This is what is known as *destructive* interference. All of the $+x$ excursions in curve A are exactly canceled by the $-x$ excursions of curve B, and thus the particle will remain motionless.

In Figure 3.4(b), the two waves A' and B' are in phase and have the same frequency. The two waves will now have an additive effect on the particle which will result in the wave shown as C' whose amplitude at all points will be the summation of the displacements of the original two curves. The waves are now said to interfere *constructively*.

The phase difference necessary for this phenomenon to exist cannot be overemphasized. For phase differences other than 0° and 180°, the waves will have a cumulative effect, but it will be a nonuniform effect on the particle and will result in a complex wave of the type discussed in Chapter 1.

If the frequencies are different, interference phenomena normally do not exist, but a related phenomenon will occur. This related phenomenon is called *beats* and occurs where there is a periodic diminution of energy, but not a total cancellation. This concept is often applied in the field of instrumental music. It is used by the musician in tuning instruments. As two frequencies approach being equal and are directly in or out of phase, the interference phenomena become audible. Periods of silence, or decreases in level, at a fixed rate equal to the differences in the frequencies produces a new tone. The new tone is called a *beat tone*. For example, two frequencies of 440 Hz and 435 Hz will produce a beat tone of 5 Hz, while 440 and 439 Hz will produce a 1-Hz beat. Thus as a player adjusts an instrument to a standard (or another instrument) the nearer to its being in tune the lower will be the beat frequency. The instruments are in tune when the beat frequency disappears. Beat frequencies, although they are a result of cancellation of two waves, create periods of silence which can be readily heard by the human ear. Such frequencies are often indistinguishable by the ear from actual source frequencies at that low rate. This fact quite frequently creates problems in the analysis of noise. Beat frequencies occurring around machines are usually true beats, and are the result of the difference of two predominant, machine-related frequencies. Only in rare instances do the low frequencies, as occur in the beat frequency phenomenon, emanate from a noise source.

The use of the interference phenomena as a means of securing any noise control of machines and related sources is an exercise in futility. Despite many popular notions, the phenomena inherently requires two sources, an exact frequency match, and an exact phase relationship. These do not as a rule exist simultaneously in environmental noise sources. Two jet engines, with infinite frequency components in all stage of phase relationships, placed back to back do not produce cancellation or silence, they only raise the resultant noise level as a function of the number of sources or as $20 \log 2 = 6$ dB. A pure tone transformer cannot be silenced by electronically feeding its signal back through a loudspeaker aimed at the transformer producing a tone 180° out of phase with the transformer signal. The noise levels go up at each corner of the transformer. Such examples have been tried and have proven to be very costly and very

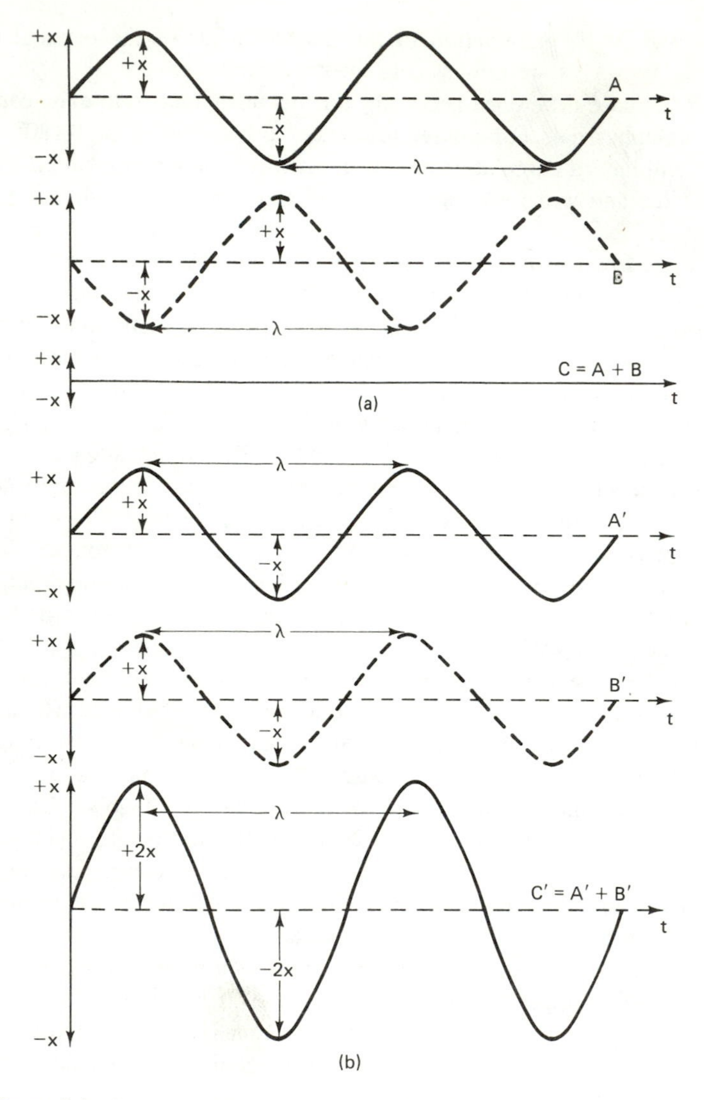

Figure 3.4 Types of interference: (a) destructive; (b) constructive; and (c) waves out of phase.

unsuccessful. The people responsible for these and other experiments attempting to use the interference phenomena have, despite their expertise in other fields, failed to recognize the basic requirements of the concept of interference. As will be seen in later discussions of noise-control devices, a few pure tone components of a noise can be eliminated or minimized by proper use of the concept.

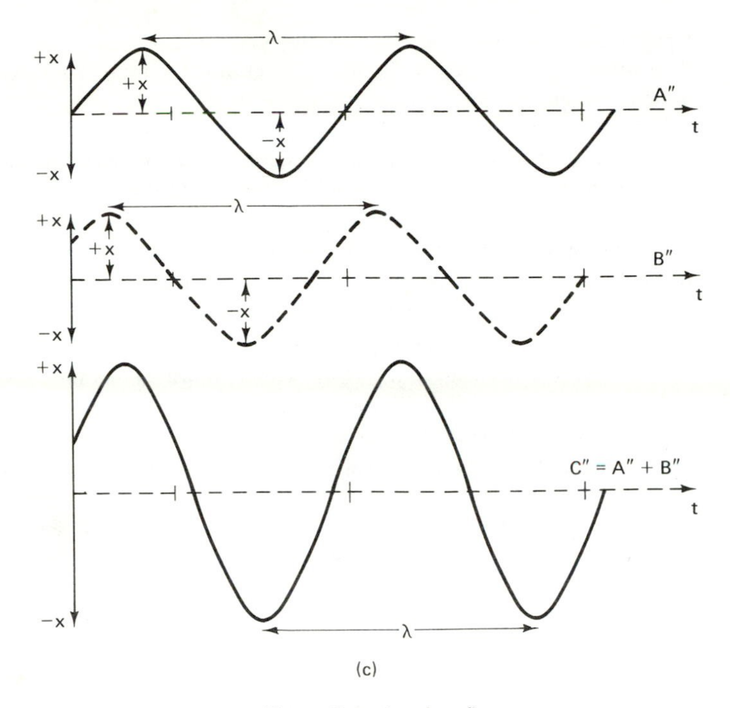

Figure 3.4 (continued)

PROBLEMS

3.1. A 6-ft tall worker in an industrial plant is normally positioned 10 ft from a machine whose main source of noise is located at a height of 3 ft above the plant floor. It is desired that a barrier 6 ft high be placed at a distance of 3 ft from the machine. Calculate, using both the modified Fresnel theory and the Fehr equation, the attenuation to be gained by the worker at 500 Hz, 1000 Hz, and 2000 Hz.

3.2. A barrier is to be erected at a distance of 50 ft from the nearest line of traffic on a highway. The barrier is to be 17 ft in height above the grade of the pavement. Trucks using the highway have their exhaust outlets at a height of 10 ft above the pavement. A house is located 200 ft from the line of traffic and its second-floor bedroom windows are at an elevation of 20 ft above the grade of the pavement.

a. Calculate the attenuation to be achieved by this barrier (using the modified Fresnel technique) at 500 Hz, 1000 Hz, and 2000 Hz.

b. What attenuation would be achieved if the windows were 10 ft above the pavement level?

3.3. In Problem 3.2(*a*) how high must the barrier be in order that a minimum of 10 dB of attenuation be achieved at 500 Hz? For that barrier what would be the attenuation at 1000 Hz and 2000 Hz?

3.4. A partial enclosure is placed 1 ft from the point of impact of a drop forge. The barrier extends 3 ft above the line of sight from the source to a receiver located a distance of 3 ft away.

a. What attenuation will be provided to the receiver at 1000 Hz?

b. How much attenuation at the same frequency will the barrier provide if the receiver moves to a distance of 10 ft away, assuming the barrier is still 3 ft above the line of sight?

3.5. A door which is 3 ft by 8 ft has a transmission loss rating of 30 dB. If the door is hung in such a way that there is a $^1/_2$-in. air space at the bottom, calculate the actual transmission loss of the assembled door.

3.6. Reconsider Problem 3.1. Suppose that the duties of the worker require him to make a close adjustment of a part by bending down so that his eyes and ears are only 2 ft above the floor for short intervals. Can he be protected by a barrier? Explain in detail using actual calculations as necessary.

3.7. Using Figure 3.4 as a model, construct on graph paper the resultant wave created by two waves which

a. Are of the same frequency and amplitude but are 45° out of phase;

b. Have a frequency ratio of 2:1, the same amplitude and are 180° out of phase;

c. Have the same frequency ratio, an amplitude ratio of 2:1, and are 180° out of phase.

3.8. Refer back to Chapter 2 and the discussion on the paths of sound rays in the hypothetical auditorium. Explain, using the interference phenomena, what audible conditions might exist at points *A*, *B*, and *C* of the diagram (Figure 2.3) under what physical conditions of the space.

4

Absorption

4.1 DEFINITION OF ABSORPTION

The concept of absorption in acoustics refers to the loss in energy occurring when a sound wave strikes and is reflected off of a given surface. The word "absorption" is frequently used by the layperson in connection with the action of a sponge when soaking up water. Such a connotation *does not apply* to acoustics. Water absorbed by the sponge can be regained by merely squeezing the sponge. Noise, as "absorbed" by acoustic tile, cannot be retrieved, it becomes lost energy (converted to heat). The concept of acoustic absorption is applicable, primarily, to interior spaces. If no walls exist, sound is totally lost as the distance from the source increases.

If we assume that a wave with a given amount of incident energy strikes a surface at a random angle, then a portion of the incident energy will be reflected back into the space where the incident energy originated and the rest of the incident energy will be transmitted into and often through the boundary material. See Figure 4-1.

Using the implied ray technique, the absorption coefficient α may be defined as

$$\alpha = 1 - \frac{\text{Reflected energy}}{\text{Incident energy}} \qquad (4\text{–}1)$$

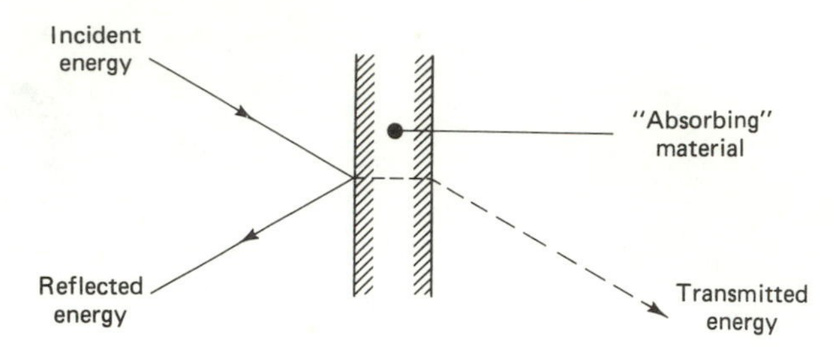

Figure 4.1 Reflection of sound energy off a plane surface.

Thus, the absorption coefficient represents that portion of the sound energy which is lost to the source space. Its value will range from 0.0 to 1.0 (that is, zero percent to 100 percent). Thus if the absorption coefficient is 0.0, no energy is lost and all of the sound stays within the source space. This implies that all of the walls are acoustically "hard" and that the reflected energy equals the incident energy. As the coefficient approaches 1.0 more and more of the energy is lost to the source space and the reflected energy becomes an increasingly smaller portion of the incident energy. The surface is said to be acoustically "soft."

Similarly, the transmission coefficient can be defined in the following manner

$$t_c = 1 - \frac{\text{Transmitted energy}}{\text{Incident energy}} \tag{4–2}$$

The total energy in the wave is represented by the sum of the absorption coefficient (α) plus the transmission coefficient (t_c) as follows

$$\alpha + t_c = 1 \tag{4–3}$$

neglecting the frictional loss (conversion to heat) which occurs within the material. This frictional loss is an infinitesimal loss at its greatest. We will see the definition contained in Equation (4–2) as the basis of our later discussion of transmission losses.

The numerical value of the absorption coefficient, as has been previously stated, for all known materials is a finite value ranging from 0.01 (1 percent) for extremely hard surfaces like polished steel or dense concrete to 0.99 for highly absorbent materials. Open windows are considered to be 100 percent absorbent. The range of practical absorption coefficients is plotted in Figure 4.2. At this point it is well to note that some manufacturers list acoustically absorbing materials of over 1.00 (that is, better than 100 percent absorbent). This is, of course, a gimmick to take advantage of the lack of basic knowledge of the concept of absorption. In the case of some products usually labeled as "unit absorbers," the material protrudes from the wall surface like a miniature box hung on the wall. The protruding surfaces are all covered with absorbing materials but the box occupies a wall area equivalent to one face. Hence more absorption per square foot appears to be available than the normal wall covering would provide. Manufacturers thus rate them as more than 100 percent absorbent. However, if these unit absorbers are installed adjacent to each other and the sound is not allowed to strike the edges, the manufacturers' claims will not be realized. For unit absorbers to be effective, they must be spaced out. If this is done, then the absorption *per square foot of wall area* drops to less than 100 percent.

The absorption coefficient is also a function of the frequency of the sound wave. Shorter wavelengths (higher frequencies) have the property of more easily penetrating the barrier and being converted into heat energy than do the longer wavelengths (low frequencies). Figure 4.2 demonstrates this concept in that the high frequencies generally have higher absorption coefficients than low frequencies.

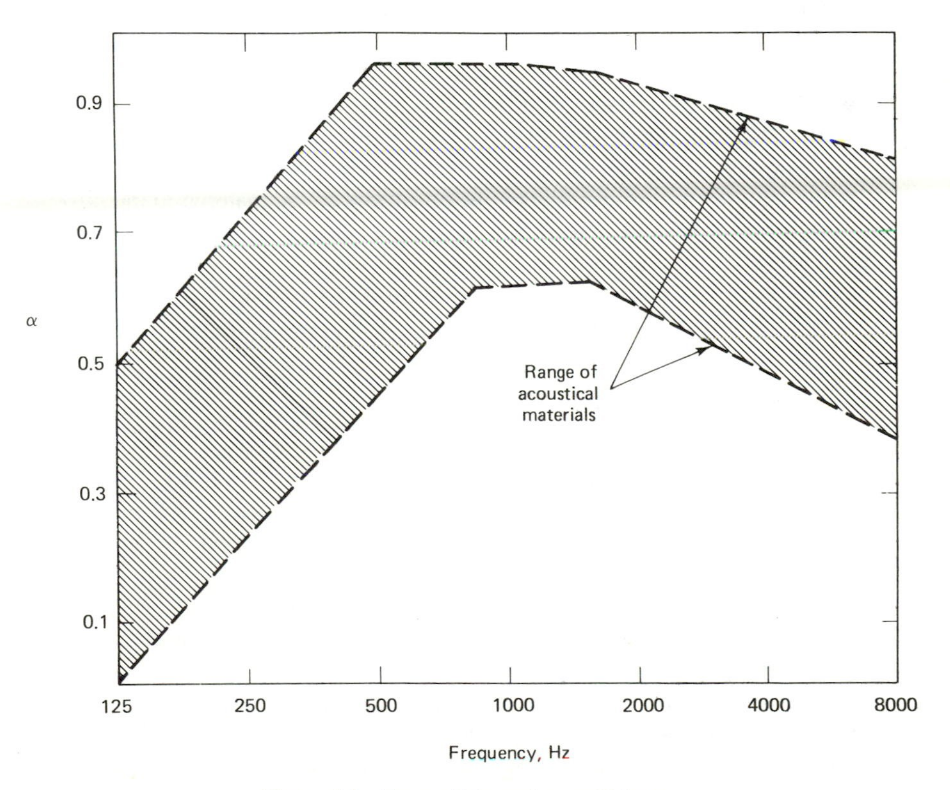

Figure 4.2 Range of absorption coefficients.

The general properties that sound absorptive materials must have to be effective include the necessity for the surface to be relatively transparent to the sound waves. As glass is transparent or translucent to light, so also are various materials to the passage of sound. The material must also provide some mechanism whereby sound energy is converted to heat energy by friction during the passage of the wave through the material.

Transparency is achieved by having the surface be either highly porous, or by use of a perforated hard material over porous material, or by covering porous material with a very thin, lightweight, flexible, air-impervious membrane. Each of these can produce the same absorptive effect, the difference being the type of atmosphere each is to be used in. All of these types of construction act as a mass-series type of acoustic reactance and therefore all of them show a decreasing transparency as the frequency increases beyond the optimum design point. Now let us take a look at the basic theory of acoustic materials.

4.2 FLOW RESISTANCE

Fibers of material and their interlocked airspaces are the frictional elements which provide the resistance to the wave motion. As the sound enters the material, its amplitude is decreased by friction as the wave attempts to move through the passages. Thus the energy of the wave is decreased. The frictional quantity is expressed by the resistance of the material to the flow of air through it and is called *flow resistance,* defined as

$$\text{Flow resistance} = \frac{\text{Pressure drop across sample}}{\text{Velocity of air passing through sample}} \qquad (4-4)$$

The units of flow resistance are defined as

$$\begin{aligned} \text{Flow resistance units} &= \frac{\text{dynes-second}}{\text{centimeters}^3} \\ &= \text{rayl} \qquad \textit{(unit definition)} \end{aligned}$$

The flow resistance must be held within limits. In most materials, it is proportional to the density of that material. Any air volume behind the material (that is, behind the surface impacted by the sound) has a significant effect on its overall flow resistance. The depth of the air space behind the material is the principal controlling factor which will add to the low frequency performance of the material. For example, the low frequency cutoff for the air space is expressed as

$$f_d = \frac{c}{2D} \tag{4–5}$$

where c is the velocity of sound in feet per second and C is the depth of the air space behind the sample in feet. Thus, the deeper the air space, the more the low frequency range will be absorbed.

Many theories have been advanced to account for the absorptivity of a material to a sound wave. As proposed by Olson, the absorption coefficient can be expressed as

$$\alpha = 1 - \left(\frac{Z - \rho c}{Z + \rho c}\right)^2 \tag{4–6}$$

where Z equals the impedance in (acoustic ohms/centimeters^2) or g-s^{-1}-cm^{-2}, ρ equals the air density in gm-cm^{-3}, and c equals the velocity of sound in cm-s^{-1}. (Note: Symbols are as used by Olson.)

In terms of the physical parameters of the material, Olson further defines the complex impedance as

$$Z = \frac{R_{\mathrm{DC}}\, d}{3} + j\left(\frac{\omega \mathrm{~d~m~} \rho}{3} - \frac{(\rho c)^2}{\omega\, pd}\right) \tag{4–7}$$

where

R_{DC} is the DC acoustical resistance of the material per unit cube

d is the thickness of the sample in centimeters

m is the ratio of the effective density of the air in the pores of the material to the open density of air

j is the imaginary coefficient $\sqrt{-1}$

P is the *porosity* or (volume of air in pores/total volume of material)

ω is the circular frequency of $2\pi f$

The acoustical resistance (R_{DC}) can be found by determining the pressure drop across a sample of material for a given volumetric flow of air through a sample. Thus

$$R_{\mathrm{DC}} = \frac{\Delta p}{U} \tag{4–8}$$

The value of m in Equation (4–7) is obscure. It depends upon the motion of the fibers in the material and of the material itself. Its value can be assumed to be 1.5. It expresses the property of the material to yield, or in other words, it is a measure of the flexibility of the material.

4.3 POROSITY EXPERIMENT

The value of the porosity (P_o) can be obtained experimentally when the following experiment is performed. Acoustic material of volume V_m is placed in a chamber whose total volume is V. The valve at the top of the apparatus shown in Fig. 4.3 is opened and the mercury column of the manometer is allowed to stabilize at h. The valve is then closed and the right side of the manometer is raised to get a value δh creating a change in pressure on the sample of $\delta p = h(980)$ dynes/cm^2. A reduction in volume of the air above the sample will occur and will be equal to $\delta V_o = \delta h(S)$ where S is the cross sectional area of the manometer. Now, if P_a is equated to the atmospheric pressure, then the porosity is determined by the relationship

$$P = \frac{P_a V_o}{V_m p_o} + 1 - \frac{V}{V_m} \tag{4–9}$$

All the terms can now be entered into the impedance equation for Z, Equation (4–7), and subsequently into the equation for the determination of the absorption coefficient, Equation (4–6).

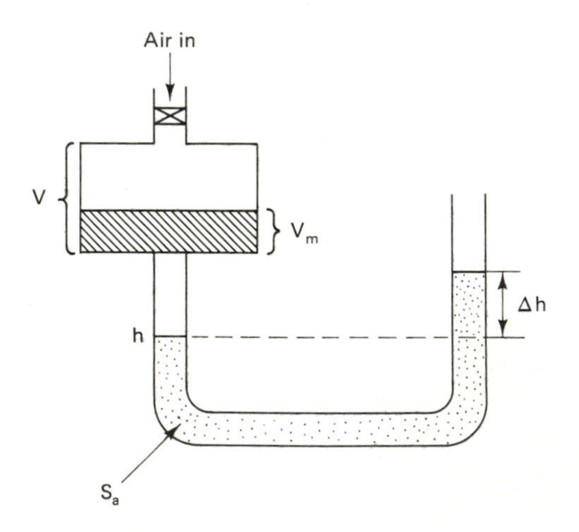

Figure 4.3 Porosity experiment.

4.4 IMPEDANCE TUBE

In addition to the theoretical discussion above, there exists a practical method of determining the absorption coefficient of various existing materials which involves the use of an apparatus called an *impedance tube*.

In the device illustrated in Figure 4.4, a sound wave at a known frequency is propagated down the tube toward the sample. It is then reflected by the test specimen in such a way as to set up standing waves in the tube.

Using a probe-tube microphone type of sensing device connected to either an oscilloscope or a sound level meter, the values of the maximum pressure (P_{max}) and the minimum pressure (P_{min}) are determined for the standing wave produced by the known frequency as emitted by the loudspeaker.

If the ratio n of the maximum to minimum pressures is known, that is, $n = P_{max}/P_{min}$, the absorption coefficient at normal incidence for the material under test is found by the relationship

$$\alpha_n = \frac{4}{n + (1/n) + 2} \tag{4–10}$$

For most practical uses the required value of the absorption coefficient will be that resulting from random rather than normal incidence on a wall, therefore the conversion from α_n to the random incidence value α_R is given by the chart in Figure 4.5.

We must repeat for emphasis that the absorption coefficient is a function of frequency. The random incidence value of the absorption coefficient is the value used in calculating the reverberation time of a given space to determine its suitability for various occupational or recreational uses.

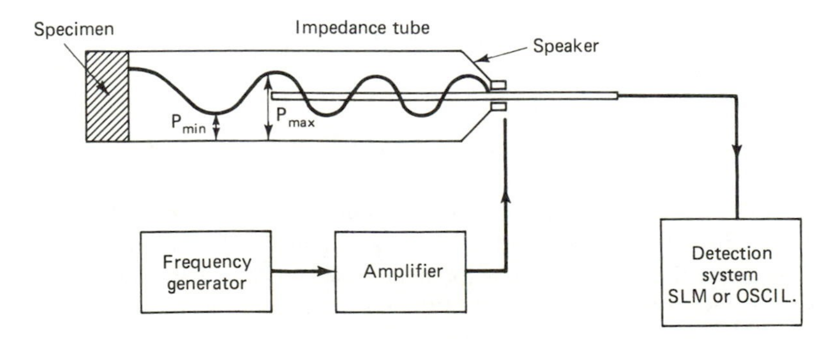

Figure 4.4 Impedance-tube experiment.

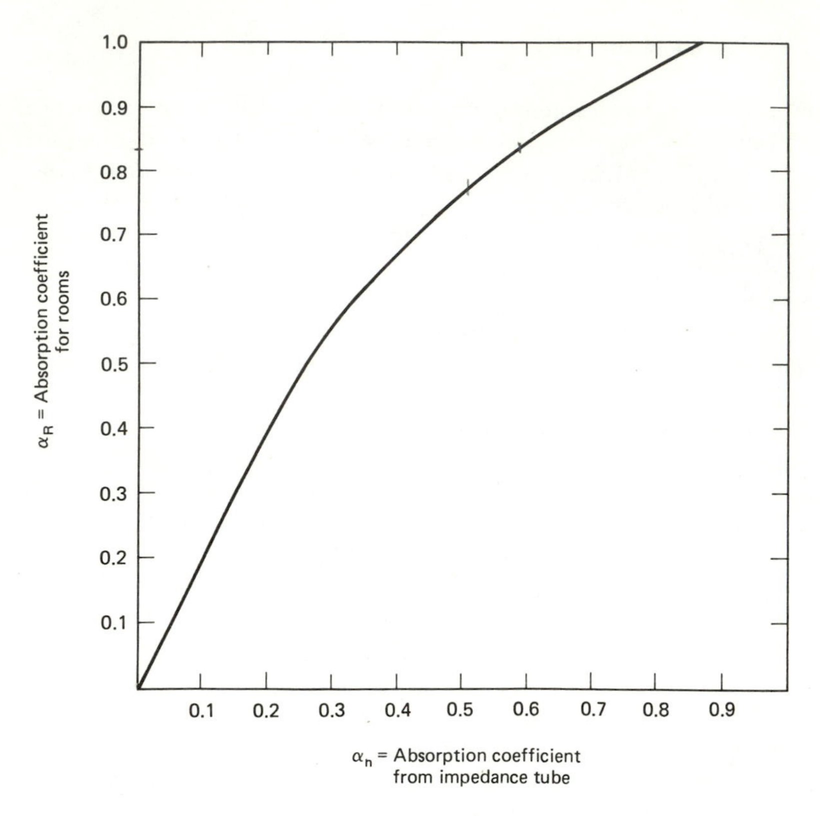

Figure 4.5 Relationship between α_R for large rooms versus α_n for the impedance tube.

4.5 THE SABIN

In Chapter 2 reverberation time was defined and the Sabin equation was expressed as

$$T_r = \frac{0.05V}{A} \text{ (s)} \tag{2–6}$$

where T_r is the time in seconds, V is the volume of the space, and A is the total absorption within the space expressed in Sabin units.

The Sabin is a unit of measure of absorption which is the product of the absorption coefficient (α) of a given material times the area S occupied by that material. Thus for an enclosed space, the total absorption A is expressed as the sum of all of the αS values within the space

$$A = \alpha_1 S_1 + \alpha_2 S_2 + \alpha_3 S_3 + \cdots + \alpha_n S_n \tag{4–11}$$

Thus, the reverberation time of a room with a given volume can be determined (or adjusted) by choosing the proper value of α for each surface, or portion thereof, multiplying it by its respective area and taking the sum of all to determine the value of A.

4.6 OPTIMUM REVERBERATION TIME

To determine the optimum reverberation time for a particular usage of a given space, it is necessary to consult a chart such as that shown in Figure 4.6. This chart is a plot of the volume of the given space versus the reverberation time with the various classes of usages plotted as a series of parametric curves. This chart has evolved through many tests conducted in many environments be determining the reactions of a "jury of peers" under various climates established by measured reverberation times. Its validity has become well established with successful applications to new situations.

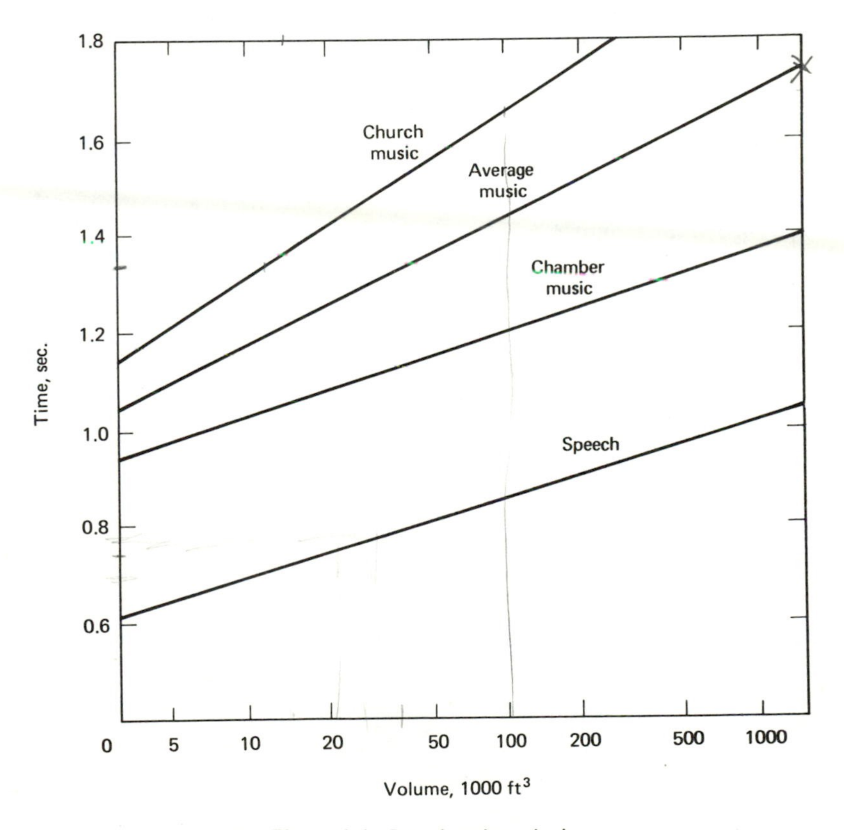

Figure 4.6 Reverberation criteria.

Having determined the optimum reverberation time as opposed to the actual reverberation time, adjustments are made in the absorption of various surface areas (αS) to achieve an optimum design. This technique will be further discussed in Chapter 12.

4.7 REVERBERATION TIME VERSUS FREQUENCY

Not unlike many other concepts in acoustics, the optimum reverberation time for a given space is a function of frequency. For a given space to be rated "acceptable" by the listener, certain frequencies must be allowed to persist for a longer period of time than others. The reason for this will be shown in Chapters 6 and 7 where the human response to various frequencies will be shown to be nonuniform as well as nonlinear.

By using research involving many polls of average people with average hearing sensitivity, we now know that low frequencies must be allowed to persist for a longer interval of time than do high frequency sounds. Thus the reverberation time for low frequency sounds must be longer than that for high frequency sounds. The proper reverberation time, therefore, must be determined not only by use of Figure 4.6 but must be adjusted by using Figure 4.7. The latter figure indicates that the optimum reverberation time for all frequencies above 500 Hz is a constant as determined from Figure 4.6. For lower frequencies, a correction factor K

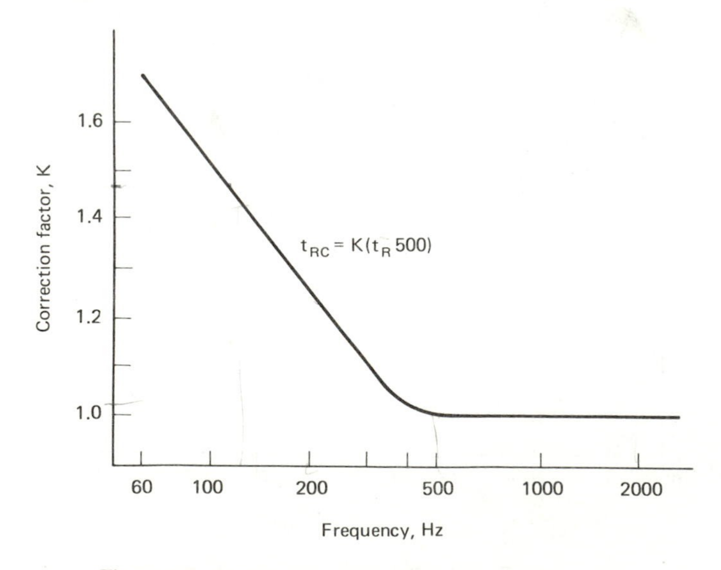

Figure 4.7 Variation in reverberation time with frequency.

must be determined from Figure 4.7. This factor, multiplied by the optimum reverberation time t_{r500}, will yield the corrected reverberation time t_{rc} at any of the lower frequencies. For example, if the optimum reverberation time as determined from Figure 4.6 is 2.0 s, then the reverberation time for that space at 200 Hz should be 1.22 × 2.0 or 2.44 s. *Failure to create this balance of reverberation time is often the result of poor auditorium design and results in an unbalanced reception of sounds by the audience.* Such conditions are often impossible to correct after the building is completed. Famous auditoriums have been condemned as a result of this condition resulting from a lack of recognition of the problem during the design stages.

4.8 ABSORPTION COEFFICIENTS OF COMMON MATERIALS

Absorption coefficients for common materials are tabulated in many textbooks. Manufacturers of materials also publish the results of extensive tests of their materials. The National Bureau of Standards publishes the results of its tests in annual bulletins.

A random sampling of the absorptive coefficients of various materials is shown in Table 4.1. Only three frequencies are indicated. Absorptive values at other frequencies can be found in more detailed tables in other references.

4.9 NOISE REDUCTION COEFFICIENT (NRC)

A term often used in describing the relative merits of acoustical materials is the *noise reduction coefficient* (NRC). The name is a paradox since it does not really refer to "noise reduction" but is a weighted average value of absorption. While this number has the value of being a single number index of the sound absorbing efficiency of a material, it is deficient in that it does not account for any special property of a material at either high or low frequencies. The NRC is defined as the mathematical average of the values of the absorption coefficients at each of the following frequencies: 250 Hz, 500 Hz, 1000 Hz, and 2000 Hz. A further deficiency in the use of this coefficient is that two materials with widely differing absorption coefficients at the same frequency can have the same NRC number. The last column of Table 4.1 demonstrates these comments concerning the NRC.

TABLE 4.1 Sound Absorption Coefficients of Common Materials

Material	Sound Absorption Coefficient			
	125 Hz	500 Hz	2000 Hz	NCR
Concrete	0.01	0.02	0.02	0.01
Concrete block				
(unpainted)	0.36	0.31	0.39	0.35
(painted)	0.10	0.06	0.09	0.05
Brick (unglazed)	0.01	0.02	0.02	0.01
Glass (window)	0.35	0.18	0.07	0.15
Plaster	0.10	0.06	0.05	0.05
Wood	0.15	0.10	0.06	0.10
Heavy drapery (18 oz/yd^2)	0.14	0.55	0.70	0.60
Heavy carpet (thick backing)	0.08	0.57	0.71	0.55
Grass (2 in. high)	0.11	0.60	0.92	0.62
Trees (thick balsam)	0.03	0.11	0.27	0.15
Water surface	0.01	0.01	0.02	0.01
Theatre audience	0.30	0.70	0.90	0.75
Acoustic tile (general-glued)*	0.10	0.80	0.78	0.75
Acoustic tile (air space)*	0.70	0.83	0.95	0.90
Fibrous glass (4 lb/ft^2)				
Hard backed				
1-in. thick	0.07	0.48	0.88	0.60
4-in. thick	0.39	0.99	0.94	0.95
Polyurethane foam				
Open cell				
1/4-in. thick	0.05	0.10	0.45	0.20
1-in. thick	0.14	0.63	0.98	0.70
2-in. thick	0.35	0.82	0.97	0.82
Hairfelt				
1/2-in. thick	0.05	0.29	0.83	0.46
1-in. thick	0.06	0.80	0.87	0.72

*Refer to manufacturers' data.

PROBLEMS

4.1. An experiment using an impedance tube results in the following set of data at a driving frequency of 1000 Hz

First P_{min} = 2.0 units   First P_{max} = 9.0 units
Second P_{min} = 1.9 units   Second P_{max} = 8.55 units

a. Calculate the average absorption coefficient at normal incidence of the acoustic material under test.

b. Determine the random incidence value of the absorption coefficient at 1000 Hz for this material.

4.2. You are asked to analyze the plans for a music room (primarily for use by choral groups) which will have a capacity of 30 people. The size of the room is 30 ft by 30 ft by 15 ft. The floor is to be covered with asphalt tile (absorption coefficient approximately the same as concrete). The walls are to be left unpainted and are to be constructed of concrete block, while the ceiling is to consist of a suspended acoustic tile (air space behind).

a. Compare the actual reverberation time of this room as designed with the optimum reverberation time for this use.

b. Is there too much absorption?

c. What would you recommend?

d. What is the room good for as presently designed?

4.3. Assume that the room in Problem 4.2 has been increased in size to 30 ft by 60 ft by 20 ft with all of the surface materials remaining the same. Analyze this new room as was suggested in Problem 4.2 answering the same questions.

4.4. Analyze the same room as described in Problem 4.3, only this time analyze it for a frequency of 125 Hz. Compare your results with the above two problems.

4.5. A conference room for a certain bank has dimensions of 30 ft by 50 ft by 15 ft. The average absorption coefficients of the room are: at 500 Hz, 0.22; at 1000 Hz, 0.30; and at 125 Hz, 0.15. Determine the reverberation time for this room at each of the above frequencies and compare the values obtained with the optimum reverberation time.

4.6. The room described in Problem 4.3 is to be used for a press conference by the President of the United States. One hundred members of the press will be present (each person has 4.5 sabins of absorption). Will this room provide a satisfactory acoustic environment? Show calculations to prove your answer.

4.7. The impedance tube of Problem 4.1 is connected to a sound level meter. At 1000 Hz consecutive readings on the meter are as follows: P_{min} = 70 dB, P_{max} = 85 dB; and P_{min} = 68 dB, P_{max} = 83 dB. Answer the same questions as you did in Problem 4.1. (*Hint:* Absolute pressures must be used to determine n.)

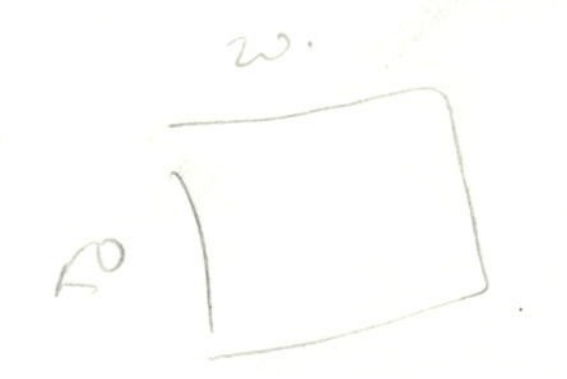

5

Transmission Loss

5.1 ENERGY LOSS

Transmission loss refers to the loss in sound energy across a wall or barrier. In Chapter 4 we saw that the transmission coefficient was defined as

$$t_c = 1 - \frac{\text{Transmitted energy}}{\text{Incident energy}} \tag{4–2}$$

Since that expression represents the proportion of the incident energy which emerges from the barrier, the energy loss across a partition will be the reciprocal of the transmission coefficient t. Also, since the *transmission loss* (TL) is usually expressed in decibels, the following relationship is a basic definition

$$\text{TL(dB)} = 10 \log \frac{1}{t_c} \tag{5–1}$$

As with absorption, the transmission loss is a function of frequency. Also, the passage of sound through a barrier is controlled by many factors dealing with the type and properties of the material present in the barrier.

Sound energy striking a hard, non-porous wall (such as dense concrete) generates compressional waves in the barrier. These waves are transmitted through the material in all directions with a resultant high energy loss due to the high stiffness or resistance of the material to the passage of a compressional wave (sound). The same energy striking a porous material (such as acoustic tile) can easily pass through the pores and thus little energy is lost and the material is said to have a low transmission loss. Acoustic tiles are useful in controlling the sound within a space where the source exists (reverberation) but by their very nature as a porous material they provide almost no barrier to the passage of sound to adjacent spaces. For example, an acoustic tile ceiling in a basement recreation room or play area will not prevent sounds from passing into sleeping quarters located above this area.

5.2 TRANSMISSION ACROSS A PLANE BOUNDARY

Before proceeding with a further discussion of transmission loss, we will now consider, in detail, the effects that exist when a plane wave impinges in a direction normal to the boundary on the interface between two media whose characteristic impedances are $Z_1 = \rho_1 c_1$ and $Z_2 = \rho_2 c_2$. See Figure 5.1.

As the incident plane wave approaches the interface or boundary it has a pressure p_i. After striking the surface, a portion of the wave is reflected with a pressure p_r and the remainder of the wave is transmitted across the boundary with a pressure p_t. During the corresponding positions, the wave will have particle velocities of u_i, u_r, and u_t, respectively.

To maintain the continuity of pressure on both sides of the interface, the following relationship must hold

$$p_t = p_i + p_r \tag{5–2}$$

And since the two media on each side of the boundary must always remain in contact with each other, the particle velocities on each side must also be equal, thus

$$u_t = u_i + u_r \tag{5–3}$$

Now, if the wave has a sinusoidal form, and if any absorbent effects of the media themselves are neglected, then

$$p_i = A_1 \sin(\omega t - k_1 x) = u_i Z_1 \tag{5–4}$$

$$p_r = B_1 \sin(\omega t + k_1 x) = -u_r Z_1 \tag{5–5}$$

and

$$p_t = A_2 \sin(\omega t - k_2 x) = u_t Z_2 \tag{5–6}$$

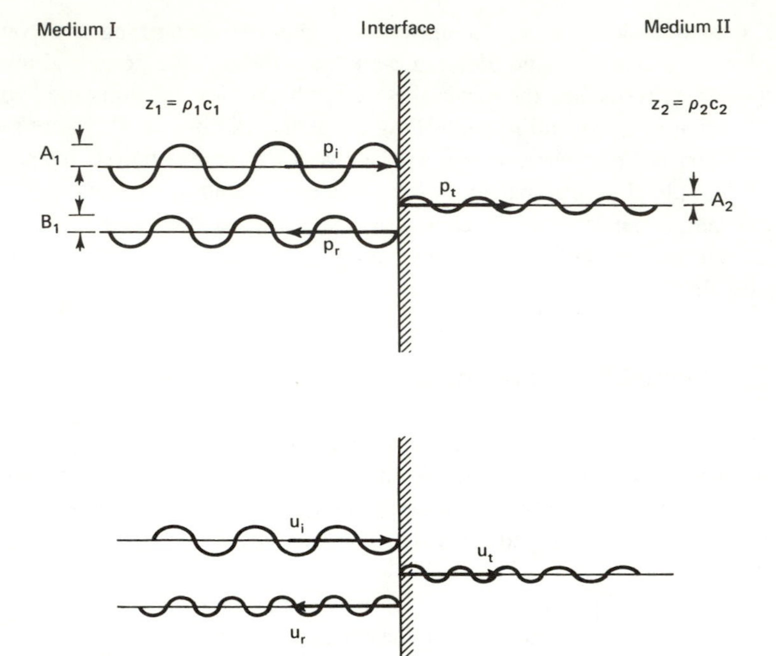

Figure 5.1 Transmission across a plane boundary.

where A_1, B_1, and A_2 represent pressure amplitudes, k_1 and k_2 equal the wave number $(2\pi/\lambda)$, and x is considered positive in the direction of the incident wave with $x = 0$ being at the interface. ω is equal to the circular frequency $2\pi f$ where f equals the oscillating frequency.

Now, inserting Equations (5–4), (5–5), and (5–6) into Equations (5–2) and (5–3), we have at $x = 0$

$$A_2 = A_1 + B_1 \tag{5–7}$$

and

$$Z_1 A_2 = Z_2(A_1 - B_1) \tag{5–8}$$

Thus

$$\frac{p_t}{p_i} = \frac{A_2}{A_1} = \frac{2Z_2}{Z_1 + Z_2} \tag{5–9}$$

and

$$\frac{p_r}{p_i} = \frac{B_1}{A_1} = \frac{Z_2 - Z_1}{Z_1 + Z_2} \tag{5–10}$$

and

$$\frac{u_r}{u_i} = \frac{-B_1}{A_1} = \frac{Z_1 - Z_2}{Z_1 + Z_2} \tag{5–11}$$

Now, defining the ratio of the energy of the reflected wave at the boundary to the energy of the incident wave at that point as the *reflection coefficient* (α_{RC}), we have from Equation (5–10)

$$\alpha_{RC} = \frac{(Z_2 - Z_1)^2}{(Z_1 + Z_2)^2} = \frac{(\rho_2 c_2 - \rho_1 c_1)^2}{(\rho_1 c_1 + \rho_2 c_2)^2} \tag{5–12}$$

The absorption coefficient defined in Chapter 4 was equal to

$$\alpha = 1 - \frac{\text{Reflected energy}}{\text{Incident energy}} \tag{4–1}$$

thus

$$\alpha = 1 - \alpha_{RC} \tag{5–13}$$

and defining the ratio of the energy of the transmitted wave to the incident energy as the *transmission coefficient* (α_{TC}), we have

$$\alpha_{TC} = \frac{4Z_1Z_2}{(Z_1 + Z_2)^2} = \frac{4\rho_1 c_1 \rho_2 c_2}{(\rho_1 c_1 + \rho_2 c_2)^2} \tag{5–14}$$

From Equation (4–2)

$$t_C = 1 - \alpha_{TC} \tag{5–15}$$

then substituting in Equation (5–2), we have

$$\text{TL} = 10 \log \frac{1}{1 - \alpha_{TC}} \tag{5–16}$$

$$\text{TL} = 10 \log \frac{1}{1 - 4Z_1Z_2/(Z_1 + Z_2)^2} \tag{5–17}$$

$$= 10 \log \frac{(\rho_1 c_1 + \rho_2 c_2)^2}{(\rho_1 c_1 + \rho_2 c_2)^2 - 4\rho_1 c_1 \rho_2 c_2} \tag{5–18}$$

The above equations demonstrate that when Z_1 and Z_2 are equal (that is, no boundary exists) the transmission coefficient α_{TC} reaches its maximum of unity—the transmission loss becomes infinite. The reflection coefficient becomes equal to zero under these conditions and thus the absorption coefficient becomes equal to unity, 100 percent absorption.

If a parallel boundary layer is considered, and a third medium is placed between the two original media, say a wall of thickness (l) with a characteristic impedance of Z_3 and a wave number k_3, then it can be shown by the same type of proof that the transmission coefficient becomes

$$\alpha_{TC} = \frac{4Z_1Z_2}{(Z_1 + Z_2)^2 \cos^2 k_3 l + (Z_3 + Z_1Z_2/Z_3)^2 \sin^2 k_3 l} \qquad (5\text{–}19)$$

These equations hold for cases where Z values are complex (resulting from phase differences) as well as those cases resulting from oblique incidence. Reference to more advanced texts is suggested for values of α_{TC} under these circumstances.

In examining Equation (5–19), it is well to note three cases of varying conditions of the value of the thickness l.

When the medium separating the two boundaries is sufficiently thin so that $k_3 l = 0$, the cosine term of Equation (5–19) goes to unity and the sine term to zero so that the equation reduces to Equation (5–14). Thus, thin sheets have virtually no effect on the transmission of sound from one space to another for the audible frequency range.

When l becomes an integral number of half wavelengths, that is, $l = n\lambda/2$ or $k_3 l = n\pi$, the equation again deteriorates to Equation (5–14). Note that the frequency dependency is again evident as in the next case.

When l is equal to an odd number of quarter wavelengths, that is, $l = [(2n - 1)\,\lambda/4]$ or $k_3 l = [(2n - 1)\,\pi/2]$, Equation (5–19) deteriorates to

$$\alpha_{TC} = \frac{4Z_1Z_2}{(Z_3 + Z_1Z_2/Z_3)^2} \qquad (5\text{–}20)$$

Thus 100 percent transmission or zero transmission loss will occur when $Z_3^2 = Z_1Z_2$.

The difficulties inherent in using the above equations for the analysis or design of existing or proposed materials for high transmission loss has been the impetus for the many techniques used in the study of the actual transmission loss of existing building materials. The more effective of these techniques will now be discussed. Practically speaking, all engineers or users of materials should be familiar with the *mass law,* the coincidence and plateau modifications of the mass law, the sound transmission class (STC), and the field or laboratory technique of measuring or expressing transmission loss.

5.3 CONTROL REGIONS

All materials have what has been classified as three control regions accounting for the transmission loss phenomena across the entire frequency spectrum. A generalized sketch of these transmission loss characteristics for each region is shown in Figure 5.2. At low frequencies there exists in all materials what is called the *stiffness-controlled* region where the resonant vibration frequencies usually are predominant. In the middle frequencies the transmission loss is controlled primarily by the mass of the material, while at high frequencies control is exercised by the location of the critical frequency.

The critical frequency f_c of any material refers to the natural frequency of resonance of that material. When any material is set into resonance at that frequency, the material's ability to transmit sound becomes infinite — there is no boundary loss. Or, stated another way, at the critical frequency the transmission loss theoretically goes to zero. This is what is meant by the *coincident effect*. Practically speaking, the transmission loss of any material drops by a considerable amount at frequencies just above

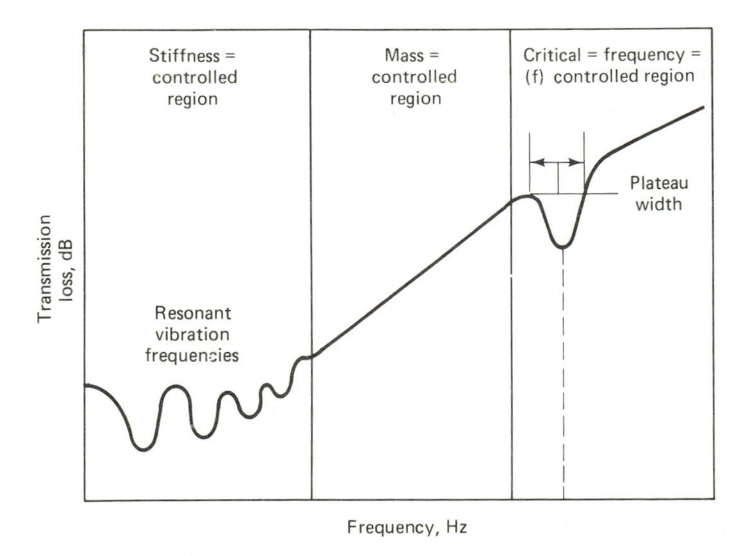

Figure 5.2 Transmission loss of homogeneous panels.

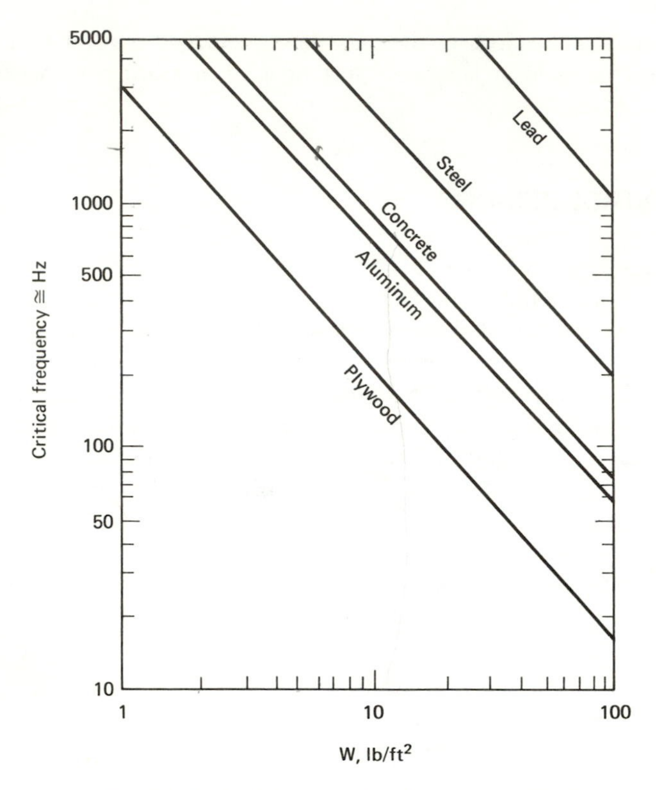

Figure 5.3 Critical frequencies of common materials.

and below the critical frequency as well. Thus a *plateau* is established where the transmission loss remains essentially constant with frequency.

In Figure 5.3, the relationship between the critical frequency f_c and the thickness of some common materials is shown. The plateau width above the critical frequency for common building materials is shown in Table 5.1.

TABLE 5.1 Critical Frequency Plateaus

Material	Plateau Width (octaves)	Frequency Ratio
Steel, aluminum, and glass	3.5	11.0
Plaster	3.0	8.0
Plywood	2.5	5.6
Lead and concrete	2.0	4.0

5.4 MASS LAW

In the mass-controlled region indicated in Figure 5.2, the transmission loss is given by the *mass law*. This is a law in which the dependency of the transmission loss on the frequency as well as the surface density is stated. *Surface density* is a unit peculiar to the subject of acoustics. Surface density is defined as the weight in pounds per square foot of a sample of material which is 1-in. thick. It is normally expressed in pounds per square foot (lb/ft^2). The value can be obtained for any material by dividing the usual density value expressed in pounds per cubic foot by 12. Thus, for concrete with a normal density of 120 lb/ft^3, the surface density will be 10 lb/ft^2 since that will be the weight of a block of the material measuring 1 ft by 1 ft by 1 in. thick. If the panel of concrete was 3-in. thick, then the net surface density W would be 3×10 or 30 lb/ft^2. If it is 6-in. thick the value would be 60 lb/ft^2. With this definition, the mass law can be plotted as shown in Figure 5.4 or written as the following

$$TL_{(dB)} = 20 \log f + 20 \log W - 33 \qquad (5\text{–}21)$$

where f is the frequency (Hz) and W is the surface density (lb/ft^2). Surface weights of common materials can be found in Table 5.2.

For practical design purposes, the calculation of the mass law is usually sufficient. The values of typical transmission loss (TL) as measured

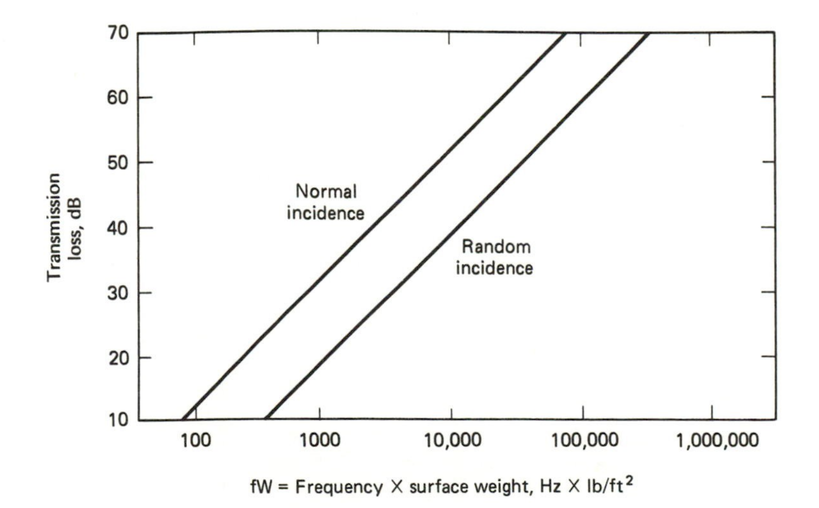

Figure 5.4 Theoretical mass law.

TABLE 5.2 Typical Surface Weights

Material	(lb/ft^2) Per Inch of Thickness	(kg/m^2) Per Centimeter of Thickness
Aluminum	14	26.6
Brick	11	21.0
Concrete (dense)	12	22.8
Cinder block (6-in. hollow)	6	11.4
Glass	13	24.7
Lead	59	112.0
Plaster (sand)	9	17.1
Plywood (fir)	3	5.7
Steel	40	76.0

in the field and tabulated in various trade journals indicate variances which come from actual tests of commercial products. Thus a more detailed analysis, combining the coincident effect with the mass law is advisable.

5.5 PROCEDURE USING COINCIDENT AND PLATEAU EFFECTS

To determine the transmission loss characteristics for a given panel of material, use the following steps:

1. On a graph of frequency versus transmission loss, construct the mass law curve in accordance with Equation (5–21).
2. From the critical frequency diagram (Figure 5.3) locate the critical frequency and plot it on the mass law curve. From this point construct a horizontal line of the indicated plateau width (Table 5.1).
3. At the end of the plateau, construct a line whose slope is upward at 9 dB per octave to the last frequency of interest. (An "octave" is a frequency range in which the higher frequency is twice the lower frequency, i.e., 500–1,000, 1,000–2,000, 2,000–4,000, etc. Each span is an octave.)

The resulting curve yields a more accurate and more practical transmission loss curve than that achieved solely from the mass law.

Figure 5.5 indicates the typical resulting curve when the above steps

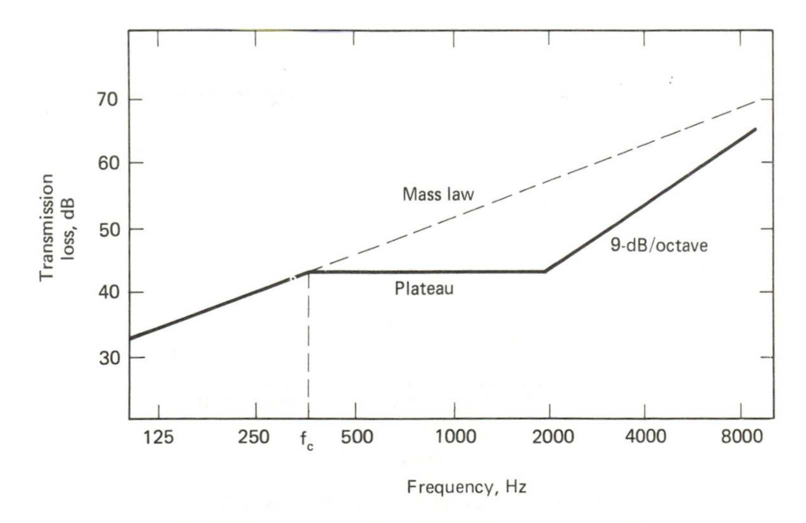

Figure 5.5 Transmission loss coincidence and plateau effects.

are followed. Figure 5.5 is plotted for a 2-in. thick slab of dense concrete. The solid curve represents the final estimated transmission loss for this material.

5.6 SOUND TRANSMISSION CLASS

A single number rating system, called the *Sound Transmission Class* (STC), has been developed to compare the transmission loss of various types of construction over a specified frequency range. This system was developed through the joint efforts of several test laboratories in conjunction with the Standards Committee of the American Society for Testing and Materials (ASTM).

As now specified by ASTM, the transmission loss of a test specimen is measured under controlled laboratory conditions at 16 one-third octave bands whose center frequencies extend from 125 to 4000 Hz. This procedure can also be used under field conditions which would then account for all leaks or flanking paths present in actual field construction.

To determine the STC of a specimen, the values of measured transmission loss are plotted against frequency. Then, either an STC curve is constructed on the same graph or a transparent overlay constructed from the graph shown in Figure 5.6 is placed over the data plot. In either case the final curve is adjusted until the following two conditions are met

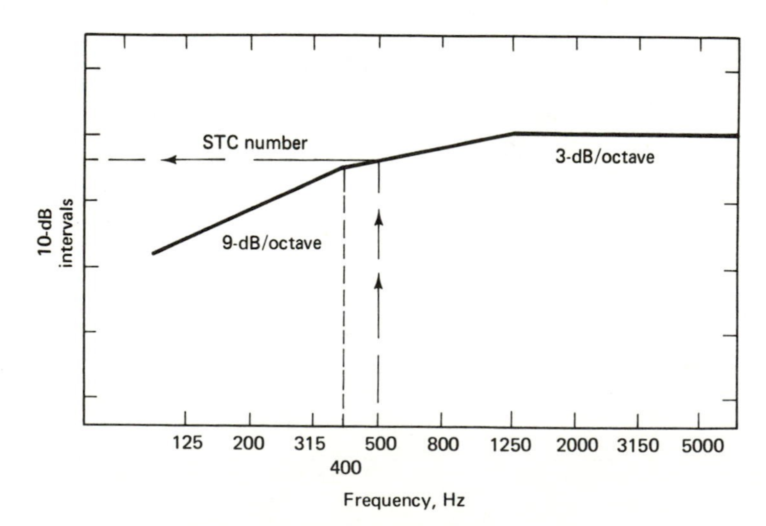

Figure 5.6 Sound Transmission Class (STC).

1. The maximum deviation from the standard curve cannot exceed 8 dB at any frequency, and
2. the average of all deviations from the standard curve cannot exceed 2 dB. Thus, for the 16 frequency bands the sum of all the deviations cannot exceed 32 dB.

When the STC standard curve is finally adjusted for the highest position which meets the above criteria, the decibel level indicated at the intersection of the curve with the frequency of 500 Hz is read from the data plot. This number becomes the STC rating of the particular test specimen.

General engineering judgment and evaluation is needed at this point. While higher STC numbers generally mean a better overall product, it is possible to have dips in the data at critical frequencies which would be detrimental to a particular application in which case the higher STC number may have to be disregarded.

5.7 FIELD TESTING—NOISE REDUCTION

In order to make a field determination of the transmission loss of an existing building partition, the following relationships are particularly pertinent. You should note, however, that confusion exists between the terms "transmission loss" (TL) and "noise reduction" (NR). The two terms are related but are not interchangeable.

Assume two rooms adjacent to a partition as shown in Figure 5.7. A source creates a sound level in Room A at a sound pressure level which

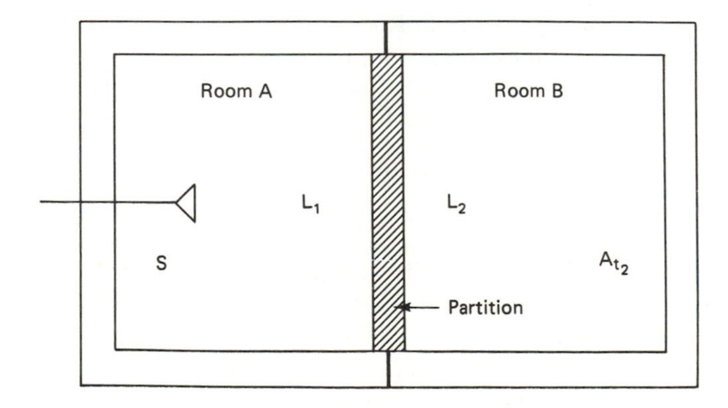

Figure 5.7 Transmission loss in a field condition.

will be designated as L_1 and a pressure p_1. A measurement of the sound pressure level in the receiving Room B is designated as L_2 and represents a pressure of p_2.

The noise reduction (NR) of the intervening partition between the rooms can then be stated as

$$NR = L_1 - L_2 = 20 \log \frac{p_1}{p_2} \tag{5–22}$$

Now, if the total absorption in the receiving room is determined to be A_{t2}, the transmission loss of the partition can be related to the noise reduction as follows

$$TL = NR - 10 \log \frac{A_{t2}}{S} \tag{5–23}$$

where S is the area of the intervening wall in square feet.

It should be clear that the pressure p_2—or the level L_2—in the receiving room will be dependent on the area of the partition wall S and the *total* absorption A_{t2} in the receiving room. This can be demonstrated by the following analysis.

The power W_1 which falls on the wall area S from the source in Room A is given by

$$W_1 = \frac{E_1 c S}{4} \tag{5–24}$$

where E_1 is the energy density in Room A.

The power W_2 transmitted from the sample or wall under test into Room B is

$$W_2 = \frac{E_1 c S}{4r} \tag{5–25}$$

where r is the loss factor due to the wall (or the transmission loss). Thus, the energy density E_2 in Room B is

$$E_2 = \frac{4W_2}{cA_{t2}} = \frac{E_1 S}{rA_{t2}} \tag{5–26}$$

so that

$$\frac{E_1}{E_2} = \frac{rA_{t2}}{S} \tag{5–27}$$

or (expressed in decibels)

$$10 \log \frac{E_1}{E_2} = \text{TL} + 10 \log \frac{A_2}{S} \tag{5–27a}$$

and if the ratio of pressures is introduced, we have

$$\text{TL} = 20 \log \frac{p_1}{p_2} - 10 \log \frac{A_2}{S} \tag{5–27b}$$

or

$$\text{TL} = 20 \log \frac{p_1}{p_2} + 10 \log \frac{S}{A_2} \tag{5–27c}$$

or

$$\text{TL} = L_1 - L_2 + 10 \log \frac{S}{A_2} \tag{5–27d}$$

which was the alternate form of Equation (5–23).

The above analysis can be performed with broad-band noise, but for more precise results, a series of frequencies from 125 to 4000 Hz should be used. Occasionally it is necessary to extend the frequency spectrum at both ends. Performing this test in the field will incorporate any flanking paths which exist by virtue of varied building techniques. The results of field tests will rarely match laboratory tests unless specific care is taken to eliminate flanking paths.

Care must be taken by the designer not to anticipate the precise results of laboratory studies when applying theoretical transmission loss data to field conditions. Even with precise and careful control of the field installation, the results can be 6 to 10 dB lower than the laboratory tests might predict. A margin of safety must be applied to this problem in a manner similar to the use of the factor of safety used by structural engineers in their design procedures. This problem will be discussed at greater length in Chapter 12.

5.8 COMPOUND PARTITIONS

The use of compound partitions made up of multiple thicknesses of various basic wall materials tightly bonded together, or the use of multiple walls, results in a greater transmission loss than the use of simple structures. Calculations of the predicted results are, however, not very precise because of the complex nature of combining the specific acoustic impedance of each component of the structure. Hence, laboratory testing of multiple wall assemblies is relied upon heavily in the design procedure.

To calculate in advance the predicted transmission loss, we must know the various resonance frequencies of each of the materials, the damping characteristics of each panel, and the coupling factor between the panels. This latter quantity is dependent upon the spacing between the walls or layers, the interdependence upon common structural support elements, and the amount of sound absorption at the various interfaces.

Compound walls will yield a greater reduction of sound from one space to another over single component walls. This increase can amount to anywhere from 6 to 20 dB. For the simpler structures of this type some calculations can be made which will serve as guidelines to the final performance of the material.

Assuming that the surface density of each of two walls is equal (that is, $W_1 = W_2$) such as in the case of two free-standing brick walls separated by an air space, then the overall transmission loss can be estimated as

$$\mathrm{TL} = 20 \log f + 20 \log W_1 + 20 \log W_2 - 35 \qquad (5\text{–}28)$$

For more exotic panel constructions, additional terms involving the surface weight can be added to the above equation. We must emphasize that this method is purely for estimating purposes and will not yield precise results.

Other factors such as critical frequency mismatching, coupling action of one layer of a multilayer, monolithic panel on another, and so on will only serve to increase the net transmission loss of this type of element. For example, the lowest resonant frequency of two closely coupled walls can be estimated by using the equation

$$f_0 = \frac{170}{x} \sqrt{\frac{1}{W_1} + \frac{1}{W_2}} \qquad (5\text{–}29)$$

where x is the separation of the walls in inches. This results in a critical frequency usually low enough to be neglected except that the transmission loss curve for the region of interest architecturally (125 to 8000 Hz) will generally rise with a slope of 9 dB per octave as compared to the general mass law slope of 6 dB per octave.

5.9 POROUS MATERIALS

Porous materials do not obey any of the mass-law characteristics as a general rule. For practical purposes, their effect can generally be predicted by relating the transmission loss (which is low in any event) to the thickness of the material. This is a linear relationship in that doubling the thickness usually results in doubling the transmission loss. Thus 4 dB of loss for a 2-in. thick blanket becomes 8 dB of transmission loss for a 4-in. thick blanket. For indepth studies on this subject, readers should refer to advanced texts and research papers. Very little practical work has been done in this area since the results are acknowledged to be minimal. Porous materials are most effective and best used in only those situations requiring absorption qualities rather than for their transmission loss characteristics.

PROBLEMS

5.1. Calculate and plot the transmission loss curve for a wall made of concrete which is 6-in. thick, 50-ft long, and 10-ft high. Use the method shown in Section 5.5.

5.2. Calculate the thickness of lead that will give a transmission loss equivalent to the 6-in. thick concrete wall of Problem 5.1. Hint: Use the mass-law equation.

5.3. Plot the transmission loss curve for the lead of Problem 5.2, including the plateau effect.

5.4. If the average noise level in a machine room is determined to be 60 dB, calculate the transmission loss of the intervening wall which is 10-ft high and 30-ft long. The average absorption coefficient for all of the surfaces of the conference room is 0.5. The room dimensions are 30 by 20 by 10 ft.

5.5. Calculate the transmission loss at the surface of a pond when a sound wave strikes it in a vertical direction and enters the water. (The ρc value for water is 1.43×10^5 rayls.)

5.6. If a sound from an underwater source strikes the water surface vertically, calculate the transmission loss to air.

6

Hearing

6.1 NEED TO STUDY

Since the human being is usually at the end of the chain as the receptor of connected phenomena related to sound propagation, it is essential that the practitioner in the field of acoustics be aware of the basic elements of the human hearing mechanism. He or she must be aware of the process of hearing as well as some of the limitations and reactions of the receiver to sounds around that individual. Complete coverage of the subject of hearing and its associated physiological and psychological effects would entail a large volume of material, and is thus not included in this text. Many references exist to which the reader may refer for a greater in-depth coverage of this topic. For the purposes of this text, the discussion merely provides a springboard—the fundamentals—from which a more detailed study can be undertaken.

Medical authorities have stated repeatedly that noise induced hearing loss is preventable. This act of prevention implies a basic knowledge on the part of engineers of the functions of the normal ear so they can prepare adequate designs to reduce the noise.

6.2 PHYSICAL DESCRIPTION OF THE EAR

A study of Figure 6.1 will reveal that the physical object called the ear can be subdivided into three parts. These consist of the outer ear, the middle ear, and the inner ear.

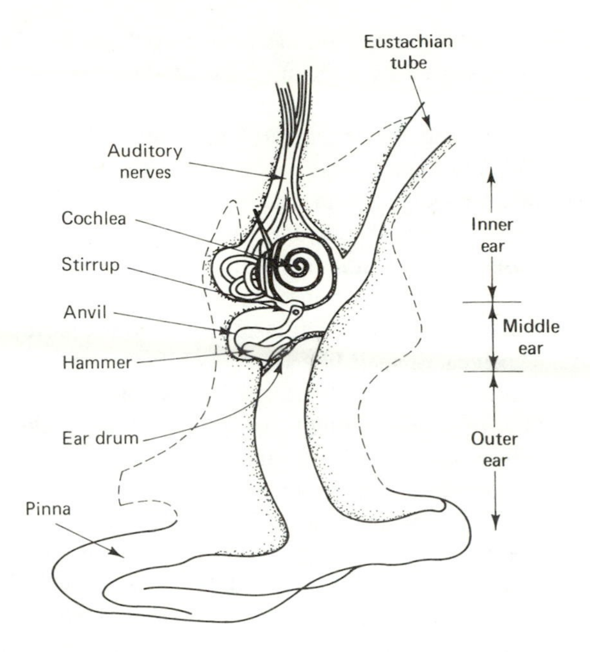

Figure 6.1 The human ear.

That part of the ear which is observed on the side of the head is referred to as the *pinna*. It serves as a collector of sound which it leads down a passage way, the ear canal, to the ear drum. Scientific authorities disagree as to the effectiveness of the outer ear as a collector of sound. Some report that the size of the outer ear has little effect on the efficiency of the collector, little ears being equally effective as large ears. Most agree, however, that this part of the ear does help in localizing the direction of the sound source as perceived by the listener. The time delay to each ear from sounds on the right or the left as well as sounds from the front and back is enhanced by the pinna.

In any event, as the sound waves move down the ear canal, they impinge on a thin membrane called the ear drum, causing it to vibrate like a drumhead. In a healthy person this membrane is extremely sensitive to pressure and frequency. One investigator, in describing the sensitivity, compared it to the case of a fly landing on the head of a bass drum. Such an impact would be sensed by the human hearing mechanism in a manner which would liken it to a clap of thunder. Any individual who has experienced the invasion of a small insect into the ear canal is aware of the intense pain that can be generated in this manner.

The action of the ear drum marks the beginning of a chain of mechanical, hydromechanical, and electrical events during the passage of the sound impulse through the hearing mechanism to the brain. The vibration of the ear drum is sensed by a chain of the smallest bones in the human body. These bones are called collectively the *ossicles* and because of their rough shape are named the "hammer," the "anvil," and the "stirrup," respectively. This mechanical system, located in the middle ear, provides both a mechanical pickup system like a vibration transducer and a mechanical multiplier or amplifier to the sound pulse. The vibration of the ear drum is sensed by the hammer (similar to the vibration pickup) and through the action of the other bones the sound pulse is amplified on the order of one million to one as it is transmitted from the stirrup to the oval window of the cochlea, or snail-like assembly, of the inner ear. This highly efficient mechanical system of bones located in the middle ear literally controls the flow of sound through the system. Also located within the middle ear is the eustachian tube running to the throat. Infection in that latter area can cause an inhibiting action on the bones of the middle ear. The purpose of the eustachian tube is believed to be that of providing a pressure balance mechanism to both sides of the ear drum. People experiencing deafness or pain when exposed to rapid changes in elevation (changes in atmospheric pressure) quickly learn to swallow rapidly thus clearing the eustachian tube and restoring the hearing to normal. Colds and other illnesses can cause blockage of the eustachian tube with the resulting decrease in hearing sensitivity. Acute infection of the middle ear inhibits the action of the ossicles and thus decreases their transmission properties with a resultant loss of hearing acuity. Problems associated with aging such as hardening of the arteries or weakening of muscular structure also account for the deterioration of the efficiency of this mechanical linkage. Speculation exists that repeated exposure to high noise levels in industry can cause mechanical fatigue which will also inhibit the action of the ossicles.

Also connected to the middle ear, as shown in Figure 6.1, are three mutually perpendicular tube-like structures filled with a body fluid which control the balance function of the body. Infections in the middle ear can be transmitted to this balance mechanism and thus cause disorientation of the individual. It is believed that disorientation which often accompanies high intensity sounds also has its origin in this area.

After the sound has passed the middle ear and is transmitted to the inner ear to the snail-like cochlea, a hydromechanical phenomenon occurs which is sensitive also to both frequency and pressure. The cochlea, if unrolled, looks like a long tube-like structure filled with a fluid and whose chamber is separated for most of its length by a membrane attached to one end. See Figure 6.2. This membrane, called the basilar membrane, has been compared to a flexible cantilevered beam because its action is very similar to one.

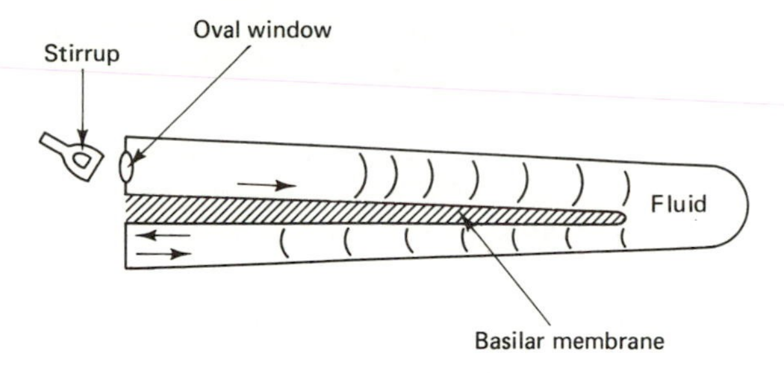

Figure 6.2 The cochlea, extended position.

When a pressure pulse from the stirrup bone is sensed at the oval window, the pulse is transmitted by the fluid down the tube of the cochlea to its narrow end, around the membrane and back to the closed end below the oval window. This pressure pulse activates a standing wave pattern within the fluid. This pattern is assumed by the membrane, thus setting the membrane, at any instant, into a fixed, wave-like deformation. The process continues as the sound pulses enter the system and the basilar membrane is continuously deforming to shapes similar to the presented wave form. Various parts of the membrane continue to displace to a greater or lesser amount depending upon the magnitude and frequency of the wave pulse.

Since the membrane is wedge-shaped in cross section, with the thin edge of the wedge being the free end, the membrane responds to various frequencies as a function of the distance from the anchored end. Thus, at the thick or attached end, the response is to low frequencies, while at the thin or free-moving end its response is to high frequency impulses.

The natural or free action of the cochlea mechanism depends to a great extent upon the viscosity of the fluid contained in it. Thus, any deterioration of the body fluids with age or illness will affect the hearing mechanism at this point. Increased viscosity or lessening of muscle tension will provide a sort of damping mechanism to which the membrane is not accustomed and thus its relative motion will be decreased.

Linked to the basilar membrane along its length are some 30,000 nerve fibers which sense the motion of the membrane and generate electric pulses which are then transmitted to the brain. Each fiber is selective with respect to frequency and amplitude, and thus in combination with the motion of the fluid and the membrane, which are also selective, transmit discrete bits of information to the hearing center in the brain. The inner ear is a very delicate structure. With long time high noise levels or sharp, high intensity sounds the fluids and nerve endings are subjected to sound pressure levels far in excess of those normally experienced. The gross effect is to cause damage to the microscopic hair cells of the inner ear and thus deterioration of the hearing acuity. The chain of mechanical, hydromechanical, and electrical signal processing is thus complete.

Beyond this point, little positive information exists as to how the brain processes information which it receives from the two ears. As many postulations exist as there are researchers in the field. Many today compare this processing action of the brain to a giant computer with all the modern memory and recall functions.

Whatever the mechanism or whatever analogy is postulated, it is evident that the human hearing mechanism is many times more diverse than any electrical or electronic sensing system yet developed.

The hearing mechanism is capable of not only receiving information, but it can catalog, store, and recall bits almost involuntarily, certainly almost instantaneously. The human being can recall and identify all sounds once heard. Adverse response mechanisms to sounds which it cannot identify can be triggered. We recognize the voice of a friend over the telephone almost immediately. A stranger is recorded as such in just as short a period of time. The ear can discriminate particular sound patterns occurring in a background of many patterns (noise). A mother can recognize the cry of her own child in the midst of the presence of noise from many other children or other sources. We can recognize the voice of an acquaintance across the room at a crowded, noisy cocktail party. The mechanic with a trained ear can hear sounds indicating particular problems in the background of machinery noise. These feats can be approached but not duplicated by existing sophisticated electronic equipment (for example, real-time sound analyzers).

In addition, the human hearing mechanism can survey information received and then trigger various bodily responses. High intensity, sharp sounds can cause fear reflexes. Automobile horns or other piercing sources such as safety signals in industry take advantage of this fact. Information gained in advanced studies in psychoacoustics leads us to the belief that many other unwanted problems can be triggered by high intensity sounds impinging on the hearing mechanism. Hypertension, high blood pressure, and adverse effects on all bodily functions are being recognized in many circumstances as a result of a direct and prolonged exposure to high noise levels.

The hearing mechanism has its own system of judging pressure or intensity which has not yet been exactly duplicated by manufactured equipment.

6.3 RANGE OF AUDIBILITY

In reviewing Figure 6.3, which is a plot of the average hearing ability of humans in pressure level (decibels) versus frequency, it is evident that the

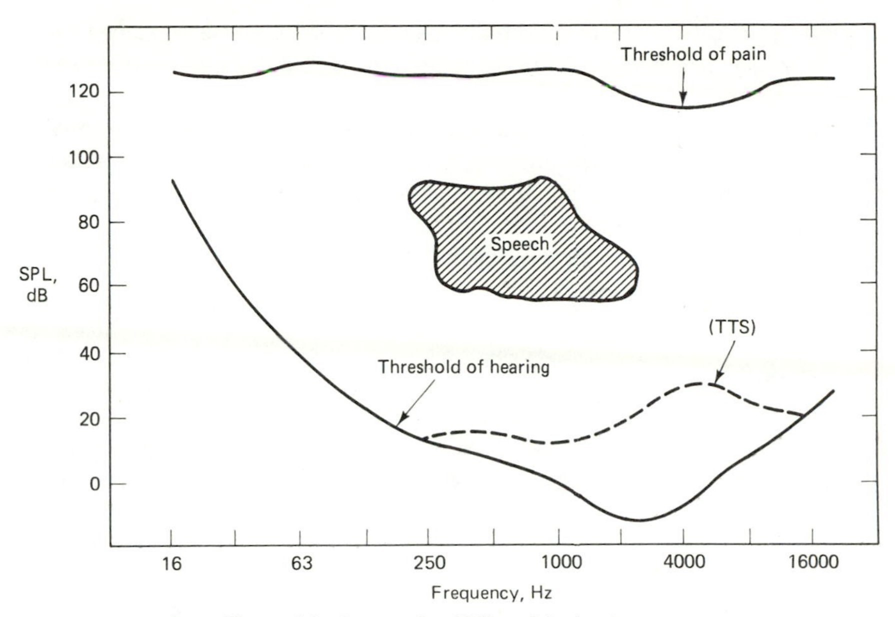

Figure 6.3 Range of audibility of the human ear.

ear is not as sensitive to low frequency sounds as it is to high frequency sounds. It is also evident that as the intensity increases, the sensitivity to all frequencies begins to equalize or become uniform. We need only to look at the shape of the threshold, or advent of the hearing stimulus, curve and compare it with the curve of response at the threshold of pain to see the flattening out effect of high intensity. Reference to these curve shapes will be made again in the discussion of the concept of loudness in the next chapter.

Note that the range of frequency response is from 20 to 20,000 Hz which compares well with the number of nerve cells described as located in the cochlea. It is interesting to note from this figure that the range of intensities and the range of frequencies of the human voice are considerably less than the range of hearing, although this is not true with other sound sources in the environment.

6.4 THRESHOLD SHIFT

Figure 6.3 also demonstrates the meaning of threshold shift as referred to in discussions of temporary or permanent hearing loss. When the ear is exposed to loud or traumatic noise levels, a defense mechanism whose origin is not fully understood is established whereby there is an apparent loss of sensitivity to new stimuli. This shift in response is named the *temporary threshold shift* (TTS) or the *permanent threshold shift* (PTS) depending upon its duration. In effect the threshold of hearing is raised (see Figure 6.3) and the sounds below that new curve are not heard until the ear recovers from the initiating stimulus. If the ear does recover, the shift is labeled temporary. If the damage to the hearing mechanism is severe and nonrecoverable, the shift is a permanent threshold shift (PTS) and results in the classification of the individual as having a permanently impaired hearing acuity.

6.5 MASKING

This term, which is not to be confused with threshold shifts, refers to the introduction of new sounds to an existing sound environment for the purpose of canceling, hiding, or drowning out (that is, masking) the existing sounds. If a new sound is introduced at a level 3 dB or more above the level of the existing noise, the new sound will be perceived by the listener as the predominant sound in the environment. Masking takes advantage of the ability of the ear to tune in what it wants to hear and tune out those sounds which are irritating or unwanted.

Masking usually refers to the situation where irritating machine noises or conversations exist and the listener desires in some manner to disregard or not hear those sounds. Masking cannot be used in those situations where the machine noise is already at the ear damage level. Masking with music is frequently used successfully in eating places to hide the noise of service wagons or dishes. Speech privacy between closely spaced individuals is guaranteed under these conditions of masking if the level is not too high. Masking with music has been used in industry, such as in bottling plants, to override the product noise and provide a more pleasing climate for the workers. Greater success has been achieved when the workers are allowed to choose the type of music. Masking with music has been used in offices designed using the open space concept. Speech privacy is again achieved.

Masking, or acoustic perfume, as it has been called, can cause irritation if the new sound level is set too high above the previously existing levels. Rock music will mask most sounds in the environment but not all individuals like this kind of music nor the level at which it is played.

From an engineering design viewpoint, masking is a method of controlling noise that should be used only as a last resort. At that time great care must be taken to specify not only the frequency content of the masking (sometimes a single frequency pure tone is used) but its sound pressure level. Otherwise the whole purpose will be defeated. It must be remembered that masking will *raise* the overall sound pressure level of the area into which such sounds are introduced.

6.6 HEARING LOSS WITH AGE

Many studies have shown that all human beings suffer a loss of hearing as part of the natural aging process. This phenomena, known medically as *presbycusis,* is not associated with environmental conditions but is due to natural changes in body chemistry, muscle tone, and so on. Widespread investigations are now going on to attempt to distinguish between loss of hearing due to aging and loss of hearing due to exposure to high environmental noise levels. This latter loss has been called *sociocusis*. The need to be able to distinguish between these two phenomena is essential to the acceptance of federal or local specifications regarding exposure to excessive noise and its effects on the hearing mechanism.

The quantification of the amount of hearing that is lost due to aging is the result of many statistical studies and all results are presently generally accepted by experts. For individual application, some tolerances must be applied depending upon the life style of the person; that is, whether they have lived in a predominantly rural or urban environment. This difference will only be a matter of a few decibels, however. Despite this problem, the generally accepted values for loss of hearing with age are shown in Table 6.1.

Unexplained factors in Table 6.1 include not only the variation in loss with frequency but the wide discrepancy in the loss when compared on the basis of sex. Interesting but inconclusive studies in the sociological ramifications of the latter differences have been performed from time to time.

In general, noise induced hearing losses, those associated with the

TABLE 6.1 Loss of Hearing with Age (in dB)

	Frequency					
	500 Hz		2000 Hz		4000 Hz	
Age	*M*	*F*	*M*	*F*	*M*	*F*
25	0	0	0	0	0	0
30	0	2	0	2	3	3
40	1	4	4	5	10	6
50	3	7	10	9	19	13
60	6	9	18	12	31	17

M = male; F = female.

individual living environment, occur first in the high frequency region. The onset of this type of hearing loss should be recognized by the individual if, after exposure to high noise levels, he or she experiences buzzing, ringing, or humming sounds in the ears. This phenomena, known as *tinnitus,* should alert a person to the very real possibility that damage to the nerve cells is occurring and that prolonged exposure can create real hearing losses.

6.7 FEDERAL STANDARDS

Studies conducted at the present time are very actively trying to establish the source levels that can cause permanent loss of hearing. Federal standards now specify the maximum exposure that may be permitted in industry related to the time of exposure of all individuals to all sounds over a 24-hr period. The limits established under the Federal Occupational Health and Safety Act (OSHA) were undergoing conflicting studies in 1981 with the aim of lowering the limits by 5 dB. The limits of occupational exposure under the 1981 standards are shown in Table 6.2.

The federal Environmental Protection Agency has proposed a lowering of the basic 8-hr exposure to 85 dB and medical authorities in the U.S. Department of Health are recommending 75 dB. Both of these groups are recommending a 3-dB equal energy increment with time rather than the 5-dB steps in the present regulations. Whatever the outcome of this controversy, it is a certainty that the levels will be reduced. When this will happen will depend on economic factors relating to the redesign or retrofitting of existing noise sources.

The time mentioned in the standard is a cumulative factor which is adjusted for exposure to varying levels which do occur during the normal work day. In other words, 95 dBA is a violation if the cumulative time of exposure is 4 or more hours but no violation exists if the cumulative exposure time is less than 4 hours.

TABLE 6.2 OSHA Limits (1978), Occupational Noise (Steady State)

Time of Exposure	Sound Pressure Level dBA
8 hr	90
4 hr	95
2 hr	100
1 hr	105
30 min	110
15 min	115

Note: for definition of dBA see Chapter 7.6

Standards for impact (short burst) noise presently permit a 140-dBA peak reading. No time factor is specified. This standard is also subject to modification based upon further study and undoubtedly will be lowered at an early date in the future. A description of the dBA scale, its origin, validity, and use is presented in the next chapter.

6.8 HEALTH EFFECTS OF NOISE

Noise has been defined as unwanted sound. It has been determined that certain categories of sound can be physically damaging to human beings. This phenomenon occurs at significantly lower levels than those indicated in the previous section. The problem deals with the various ways noise intrudes on the everyday activities of the listener. The intrusion leads to frustration, particularly if the listener has no control over the source.

Frustration leads to physiological stress where the body reacts by releasing adrenaline. The blood pressure increases and muscles tense. Thus, this evidence strongly suggests that there is a link between the unwanted sounds and cardiovascular problems, particularly hypertension.

Reports from the Council on Environmental Quality and the federal Environmental Protection Agency suggest that more than half the population of the United States are exposed to noise levels that may interfere with everyday activities. Activities cited in these reports include sleep interference, communications interference, and effects on job performance.

TABLE 6.3 Effects of Noise on Public Health

Body Function	Effects
Hearing	Permanent damage Not solely occupational Discomfort—social isolation Hinders educational process Language skills hindered
Heart	Faster pulse Increased blood pressure Increased respiration rate Increased adrenalin Increased hormone production
Other	Digestive changes (ulcers) Muscles tense Perspiration
Sleep	Quantity and quality affected Elderly and sickly are most sensitive
Newborn	Link to low birth weights and abnormalities

TABLE 6.4 Annoyance Factors in Noise Signal

Loudness
Suddenness (expected or not)
Degree of association to fear
Duration (usually the shorter the better)
Intrusion into the background
Time of day or night
Information content (necessary or not)
Variability (steady, rhythmic, or impulsive)
Spectral composition (low frequency or high frequency)
Presence of pure tones

It is generally considered to be a myth that people become totally accustomed to noise. Individuals may become used to low-level noise, but evidence indicates that the human body will make automatic and unconscious responses when exposed to either sudden sounds or loud sounds. In effect, the body reacts to noise as if it were a threat or warning, even though most noise does not mean danger. Hence the physiological stress patterns mentioned above become established.

To protect against the long-term effects of noise, the Environmental Protection Agency has recommended the following guidelines or goals to be achieved. To protect against activity interference and annoyance, it is suggested that a goal of L_{dn} = 55 dBA or less for outdoor noises be achieved. It is also suggested that an L_{dn} = 45 dBA be the goal for indoor noises. Both of the L_{dn} (day-night average noise level) values are recommended as yearly average values, although they may be taken as a guide for shorter periods of time.

The effects of noise on public health as well as annoyance factors in the noise signals can best be summarized by referring to Tables 6.3 and 6.4.

PROBLEMS

6.1. Explain the following phenomena, referring to Table 6.4:

- **a.** Flyover aircraft noise is more objectionable in rural areas than in cities.
- **b.** Truck noise on superhighways is more noticeable than automobile traffic.
- **c.** Your neighbor's outdoor parties are more objectionable than yours.

6.2. Referring to Table 6.3, describe any personal experience which you may have had that can be explained by the effects indicated.

6.3. Explain why the noise from a party of 30 is usually no greater than that from a party of 12.

6.4. Referring to Table 6.1, describe a sociological situation which may be explained by the difference in response of the male as opposed to the female.

7

Loudness and Perceived Noise

7.1 LOUDNESS

Loudness is a subjective response of human beings to sounds or noises around them. While loudness is proportional to the intensity of the noise, it cannot be expressly measured by the use of a sound level meter. However, the sound level meter can be used as an indicator to relate fixed decibel levels (dBA) to the general response characteristics of a population. All methods of measuring loudness depend upon a comparison by a jury of peers of a specific noise level with other known sound levels at prespecified frequencies or narrow bands of noise. Thus loudness is judged as being "twice as loud" or "four times as loud" as some known reference level used for comparison.

Several methods for rating the "loudness" of sounds are in use throughout the world. Each of these will provide in one manner or another a basic description of the effect of the sound as presumed to be heard and judged by the listener. Some of these systems have been incorporated into governmental control regulations for various types of sources. Thus there exist proposed individual systems for judging community noise, traffic noise, aircraft noise, machinery noise, and so on. All of these systems can in fact be tied together. In this chapter, the various systems for rating loudness will be developed in a semihistorical sequence. Evidence of the equality of the various systems will also be demonstrated.

7.2 EQUAL LOUDNESS CURVES

The classical early work in the study of loudness by Fletcher and Munson of the Bell Laboratories in connection with their research into the quality of sound reproduction of telephones resulted in the development of a series of what have been called Equal Loudness Curves. See Figure 7.1. The contours or general shape of these curves can be compared with the characteristic threshold of hearing curve shown previously in Chapter 6.

The construction of the equal loudness curves and their later modification is based upon statistical averages obtained from many panels of randomly selected individuals. A comparison of sounds at various frequencies as to their loudness was made by comparing the given sound to the sound of a 1000-Hz tone at a known sound pressure level.

A reference tone of 1000 Hz was chosen because it is at that frequency that the threshold of hearing has an indicated sound pressure level of zero

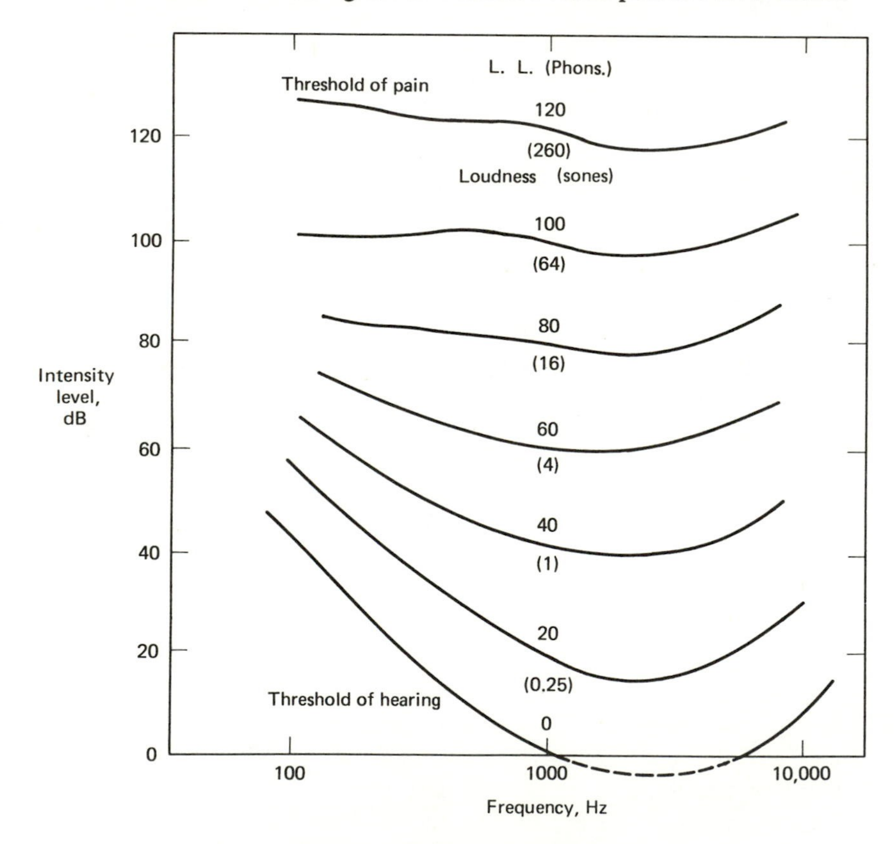

Figure 7.1 Equal Loudness Curves.

(or a close approximation to zero). It is thus the frequency at which the ear is most acute. The additional units of phons and sones are introduced to be used in determining scales of loudness level and loudness.

A *phon* is defined as that level (expressed in decibels) of a 1000-Hz tone which sounds equally as loud to the observer as the noise being compared or measured subjectively. Thus if the noise being measured sounds to the observer equally as loud as a test signal at 1000 Hz which has a sound pressure level of 40 dB, then the noise is said to have a loudness level (LL) of 40 phons. Note that if we desire to add the accumulated noise of several sources, we cannot do it by simply adding phons any more than we can directly add decibels as indicated in Chapter 1. We must use the unit of the sone.

Sone units are based upon the following definition: A loudness of one *sone* is equal to a loudness level of 40 phons. Sone units may be added arithmetically.

If So is set equal to the loudness level in sones, and Ph is set equal to the loudness level in phons, the relationship between the two can then be expressed by

$$\log So = 0.03\,Ph - 1.2 \tag{7–1}$$

An important characteristic of the loudness (sones) scale is that for a sound to be judged twice as loud as another, there must be an increase in sound pressure level of 10 dB. See Figure 7.2. For a fourfold increase in loudness, the sound pressure level must increase by 20 dB. The reader should compare this characteristic, which is inherent in the human ear, to pressure and intensity increases. In those units, a doubling of intensity is represented by a 3-dB increase in level while a doubling of the pressure is represented by a 6-dB increase. The ear with its judgment capabilities does not behave like a sound level meter in this respect.

When loudness levels in phons are to be added, they must first be each converted to sones by use of Figure 7.2 and then added by use of the equation

$$So_t = So_m + 0.3\,(\Sigma So - So_m) \tag{7–2}$$

where

So_t is the total value of sones (at different frequencies)

So_m is the maximum value in sones of any given frequency

$\Sigma\, S$ is the sum of all values of sones to be added

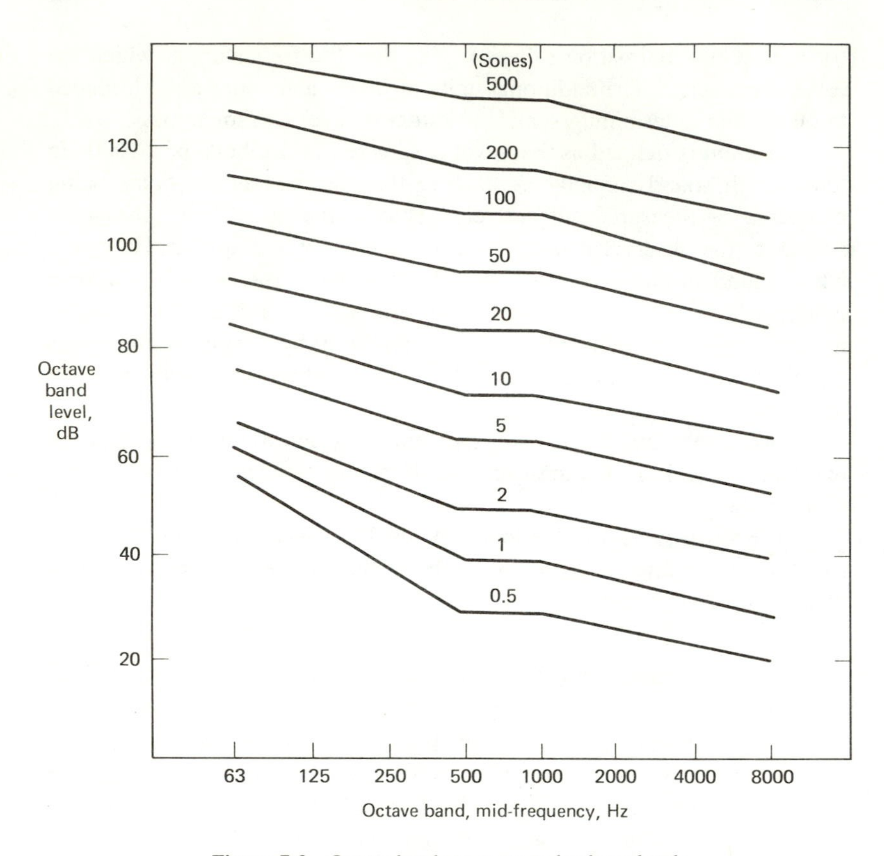

Figure 7.2 Octave bands to sones to loudness levels.

The total value of the sones, So_t, is then converted back to the loudness level in phons for the summary value by the equation

$$LL_{(\text{phons})} = 40 + 33.3 \log So_t \qquad (7\text{–}3)$$

Thus, for example, 70 phons plus 70 phons does not equal 140 phons. Rather each 70 phon loudness level equals 8 sones of loudness. Thus the sum is 16 sones which converts back to 80 phons. This would indicate that when two equal sources expressed in terms of loudness level are added the result is an increase of 10 phons or 10 dB. Again the reader is reminded of the effect of doubling pressure or intensity.

Since no generally accepted meters exist to directly measure loudness level, the greatest use of this system is in the conversion of measured octave-band sound pressure levels into a single loudness level expressed in phons. This is illustrated in the following example using conversion units from Figure 7.2.

Octave Band Midfrequency (Hz)	Measured Source Sound Pressure Level (dB)	Conversion to Loudness (Sones)
63	85	10
125	82	14
250	78	14
500	72	10
1000	65	8
2000	60	5
4000	55	4
8000	50	3
		Total 68

Now since total loudness is

$$\text{sones} = So_t = So_m + 0.3\,(\Sigma So - So_m)$$

where

$$So_m = 14$$

and (7–2)

$$\Sigma So = 68$$

then

$$So_t = 14 + 0.3\,(68 - 14) = 30.2 \text{ sones}$$

and to get the loudness level in phons

$$\begin{aligned} \text{LL} &= 40 + 33.3 \log So_t \\ &= 40 + 33.3\,(1.480) \qquad (7\text{–}3) \\ &= 89.3 \text{ phons} \end{aligned}$$

Thus from the sound pressure level spectrum for this source it is indicated that this source "sounds as loud" as a 1000-Hz tone whose sound pressure level would be measured at a value of 89.3 dB.

7.3 PERCEIVED NOISE (PNdB OR PNL)

During the past 10 years many people involved in acoustic research, felt that the loudness concept of phons and sones was causing confusion (in the United States more than any other country), when the units were compared to the decibel, and worked on developing a system of computing another unit to express the same concept. This work evolved into what is known as the *perceived noise concept* using as a unit the *perceived noise decibel* which in abbreviated form is expressed as PNdB and is often called the *perceived noise level* (PNL).

This system is based on using the sound pressure level of a band of noise over the frequency range of 910 to 1090 Hz that "sounds as noisy" as the sound or noise under comparison. Note that the single frequency comparison is not used.

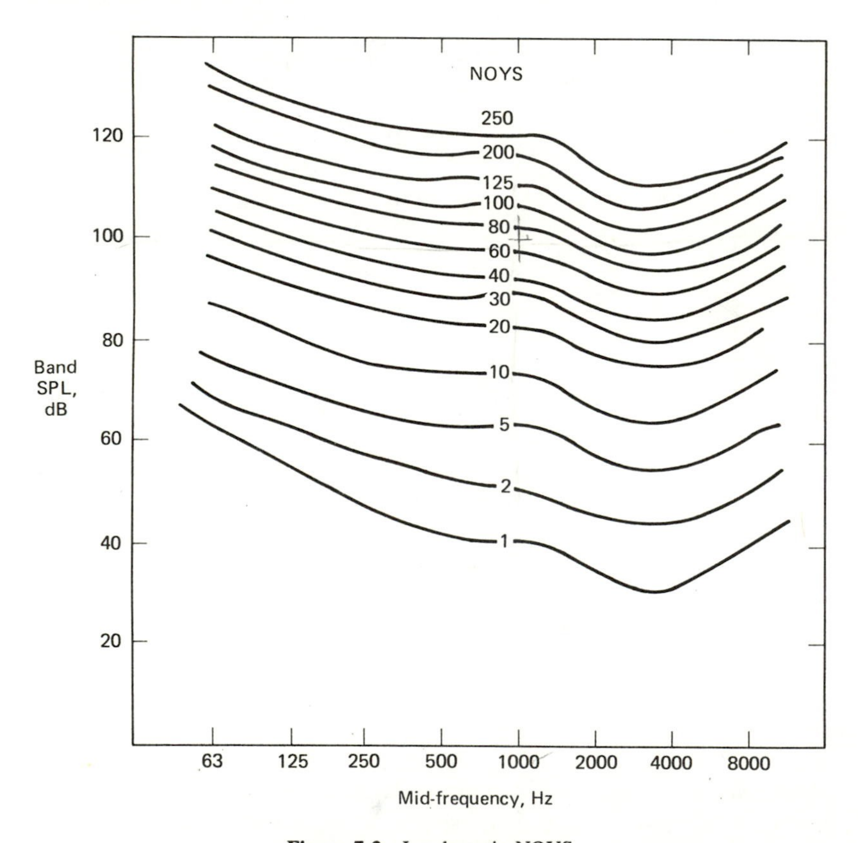

Figure 7.3 Loudness in NOYS.

In place of the equal loudness curves a set of equal annoyance contours was developed using a unit called the NOYS. These curves are shown in Figure 7.3 and in fact so closely resemble the equal loudness curves of Figure 7.1 that some critics claim that the curves are in fact equal within measurement tolerances. This has raised the speculation that perceived noise decibels (PNdB) are in fact loudness levels (phons) in disguise. Whatever the case, PNdB appears to have more public acceptance than the older phon system. This is particularly true in the United States but not in the countries of Europe where the phon and sone system remains firmly entrenched. It is well to note that, in Figure 7.3, the NOYS contour equal to 1 passes through 40 dB at the frequency of 1000 Hz.

The system used for calculating the PNdB of a noise involves the same procedure as that used for calculating the loudness level. The steps are

1. Convert the sound pressure level per band in decibels to noisiness in NOYS using Figure 7.3.
2. Sum the values of the noisiness, thus

$$\Sigma N = N_1 + N_2 + N_3 + \cdots + N_n$$

3. Identify the maximum value of $N = N_m$.
4. The total noisiness (NOYS) is then given by

$$N_t = N_m + 0.3\,(\Sigma N - N_m) \qquad (7\text{–}4)$$

5. The perceived noise level (PNL) is given by

$$\text{PNL} = 40 + 33.3 \log N_t = \text{PNdB} \qquad (7\text{–}5)$$

Figure 7.4, shown on page 92, is a nomograph which can be used to convert NOYS to PNdB.

The similarity of the two systems (loudness level and perceived noise) is inevitable since both systems are attempting to reduce a subjective human evaluation of loudness or annoyance (perhaps equal or interchangeable words in this respect) into an absolute unit.

Both systems have the same deficiency in that neither of the end result units—phons *or* PNdB—*can be directly read from any existing sound measuring equipment*. A computer is required for the conversion of any direct readings into either of these units.

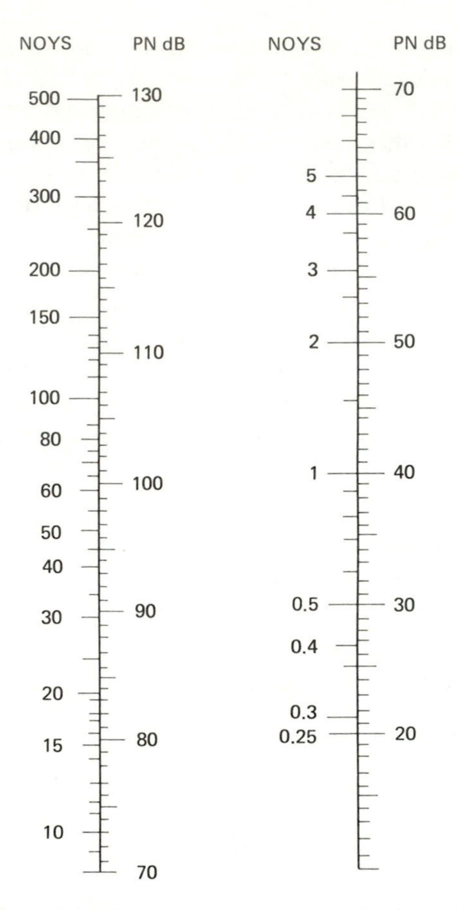

Figure 7.4 Nomograph for converting NOYS to PNdB.

7.4 EFFECTIVE PERCEIVED NOISE (EPNdB OR EPNL)

A word at this point should be mentioned about a modification of the perceived noise system called *effective perceived noise* which use a unit called EPNdB. This system modifies the above procedure by providing for empirical corrections to account for the presence of pure tones in the source and for the time exposure to the source. If the time of exposure is very short and the pure tone content is low, this latter system deteriorates into the simple perceived noise level (PNL).

The system was developed for the aircraft flyover type of noise and is under severe criticism from many practitioners not only because the values cannot be directly read from meters but because of the empirical and somewhat arbitrary nature of the correction factors and the inherent requirement that complex instrumentation and data processing equipment be used both for the time factor and for the determination of the existence of significant pure tones.

7.5 NOISE EXPOSURE FORECAST (NEF)

A further complex development of the EPNdB system is called *noise exposure forecasting* (NEF). It is used for predicting the response of people on the ground to the noise of aircraft landing or taking off from airports. Additional parameters built in to the NEF are items such as type of plane, whether landing is within the daytime or nighttime hours, what the particular landing procedure is, the time of duration of the flight over a given spot on the ground, and so on. In addition, there is a gross factor of 88 built into the equation such that the resulting number calculated for the NEF cannot possibly be confused with any sound readings taken by direct measurement. Suffice to say that this system has not had universal acceptance. The reason is self-evident when the equation for NEF is examined. Thus

$$\text{NEF} = \text{EPNL} + 10 \log (F_d + 16.7\, F_n) - 88 \qquad (7\text{–}6)$$

where F_d is the number of flights into or out of an airport between 7 AM and 10 PM, and F_n is the number of similar flights between 10 PM and 7AM on a given day. Noise of nighttime flights is given a somewhat arbitrary weighting of 16.7 to 1 over daytime flights to account for their greater apparent perceived noisiness at night when the general background noise is usually 10 or more decibels lower than in the daytime. No account is made for variances in ground elevation above runway elevations that do exist in the environs of many airports. The NEF system is severely idealistic and should be regarded as such.

Despite the above comments, the NEF system is used for evaluating the predicted noise exposure of populations located in close proximity to airports by the Federal Aviation Administration (FAA) and the U.S. Department of Housing and Urban Development (HUD). Both of these groups consider land with an NEF number of greater than 40 to be land unsuitable for habitation, between 30 and 40 to be areas of complaint, and only for numbers less than 25 to be virtually unaffected areas. Unfortunately, because of the potential for large discrepancies in the calculation of the NEF number for a given area, these numbers, while treated rigidly by the agencies, are subject to great controversy when compared to actual sound readings taken in these areas. Another system does appear to be coming into greater use and is discussed in the next section.

7.6 THE A SCALE

This standard scale appears on all sound level meters. Measurements on this scale are read in decibels (dBA) and are a measure of human response. This, in effect, gives a decibel reading which does correlate to judgments of loudness or perceived noise from broad-band noises which occur in the environment of human beings.

With reference to Figure 7.5, note that the bottom half of the figure represents the same threshold of hearing curve that was discussed in Chapter 6. Examination of this curve indicates that the normal ear rejects (does not hear) sounds at low frequencies according to a known and measurable curve. The top half of Figure 7.5 indicates the results of the introduction of an electronic weighting network to the sound level meter which filters and rejects low frequency sounds in an analogous manner to the normal performance of the ear. Readings from this scale of the sound level meter (the A scale) will then produce a single number (in decibels) which simulates the reception of that sound by the human ear. Since this reading is taken on the setting designated on the meter as the A scale, the reading is designated as dBA.

Many professional consultants in the field of accoustics have collaborated to review their field measurements of the dBA of various environmental noise sources in comparison with calculated values of perceived noise from these same sources and have produced what is now, by consensus, considered to be a valid estimate of the perceived noise by the simple equation

$$\text{PNdB} = \text{dBA} + 13 \tag{7–7}$$

This generally results in a figure for PNdB which is within 1 dB of the calculated value. Considering measurement error tolerances this is almost perfect technical agreement. For this reason as well as the fact that dBA is an immediately measurable quantity on the site, and for other reasons which will be discussed in the section on instrumentation, the use of the dBA as a measure of sound exposure is considered to have great validity when determining occupational safety regulations (see Chapter 6),

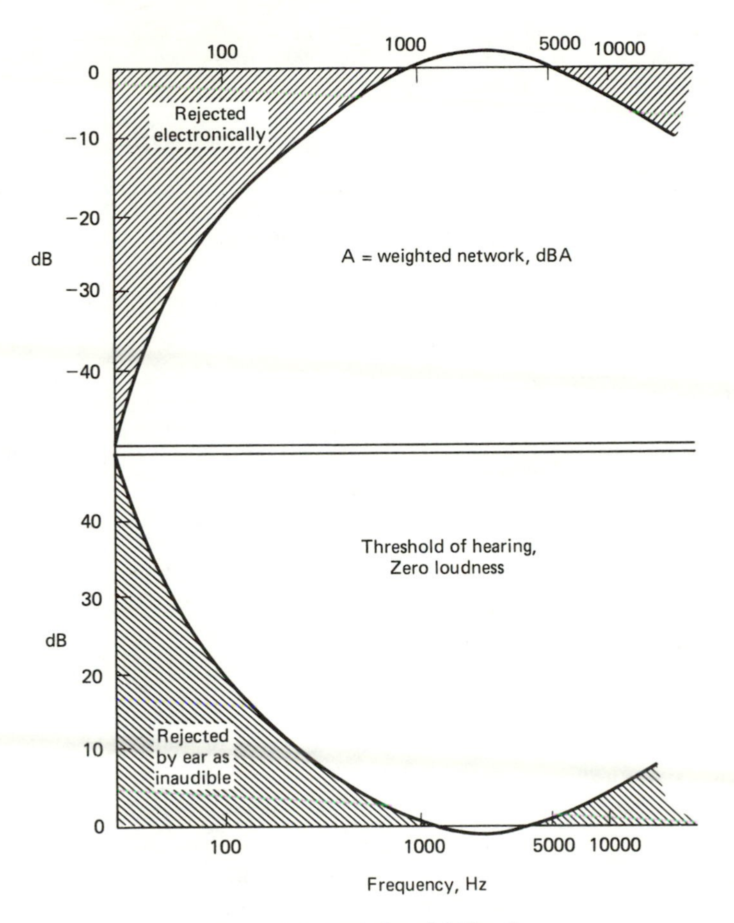

Figure 7.5 Deviation of "A" scale.

community noise response, motor vehicle regulations, stationary noise description, and general public health and safety criteria. It is being considered, although reluctantly, as a replacement of the NEF predicting and controlling aircraft noise as well.

7.7 CONVERSION TO A WEIGHTED SCALE

A sound spectrum determined in octave-band levels may be converted to a single equivalent A scale reading by the following process. The level in each respective band must be corrected by each of the following numbers:

Midfrequency octave band	31.5	63	125	250	500	1000	2000	4000	8000
A scale correction	−39.2	−26.1	−16	−8.6	−3.3	0	+1.4	+1.8	+1.9

This will yield an A scale spectrum whose numbers can then be added by the process for adding decibels described in Chapter 1.

Using the same spectrum as used in the example of Section 7.2, the process is illustrated as follows

Octave Band Midfrequency (Hz)	Measured Source Sound Pressure Level (dB)	A Correction (dB)	A Spectrum (dB)
63	85	−26.1	58.9
125	82	−16	66.0
250	78	− 8.6	69.4
500	72	− 3.3	68.7
1000	65	0	65
2000	60	+ 1.4	61.4
4000	55	+ 1.8	56.8
8000	50	+ 1.9	51.9

By the method of decibel addition, the last column will yield a single A scale value for this spectrum of 74 dBA. The perceived noise would then be approximately 87 PNdB which compares favorably with the previously calculated value of 89.3 PNdB.

While the A scale system of measurement will yield direct values of sound pressure which can be used as a measure of human response, it can, as shown above, be related to perceived noise and, with minor assumptions, to NEF. As if this were not sufficient confusion, the engineer dealing with noise problems must also be aware of the various means of rating noise which appear in contemporary literature and which seems to create a dense jungle of acoustic terminology. Some of the more prominent of these will be touched upon in the following sections of this chapter.

7.8 EQUIVALENT LEVEL (L_{eq})

An energy based form of rating noises is represented by

$$L_{eq}\ (\mathrm{dB}) = 10 \log \left\{ \int_{t_1}^{t_2} p_A^2\, dt/[p_o^2\,(t_2 - t_1)] \right\} \qquad (7\text{–}8)$$

where $P^2{}_A$ is the mean square A weighted sound pressure

$P^2{}_O$ is the reference sound pressure = 2×10^{-5} N/m^2

t_1 and t_2 are the time periods of integration

7.9 DAY-NIGHT AVERAGE LEVEL (L_{dn})

In those cases where the daytime and nightime levels are in considerable variance, the average level for a 24-hr period can be found by

$$L_{dn}(\text{dB}) = 10 \log \left[\left(\int_{0700}^{2200} p_A^2 \, dt + 10 \int_{2200}^{0700} p_A^2 dt \right) \div (24 \, p_o^2) \right] \quad (7\text{–}9)$$

where time is measured in hours and the limits of integration specify day versus night.

7.10 COMMUNITY NOISE EQUIVALENT LEVEL (CNEL)

This system is similar to L_{dn} except that it breaks the 24-hr period into three segments giving the evening hours a higher weighting than daytime hours by a factor of 3 to 1. Thus

$$\text{CNEL (dB)} = 10 \log \left[\left(\int_{0700}^{1900} p_A^2 \, dt + 3 \int_{1900}^{2200} p_A^2 \, dt + 10 \int_{2200}^{0700} p_A^2 \, dt \right) \div 24 p_o^2 \right] \quad (7\text{–}10)$$

7.11 COMPOSITE NOISE RATING (FOR AIRCRAFT NOISE) (CNR)

Since the number of flights both during the daytime and at night and thus the number of exposures of the population to aircraft noise affects the response of that population to the noise, the following system accounts for this factor.

$$\text{CNR (dB)} = \text{PNL}_{\text{max}} + 10 \log (F_d + 16.7 F_n) - 12 \quad (7\text{–}11)$$

where

PNL_{max} is the highest level of transient noise obtained in any 0.5-s time period,

F_d is the number of noisy events in daytime hours (7 AM to 10 PM)

F_n is the number during the night

7.12 NOISE AND NUMBER INDEX (NNI)

This system, which originated in England and is in use in several countries in Europe, is identical to the CNR equation [Equation (7–9)] the coefficient of the second term is 15 rather than 10 thus giving greater weight to the number of events than is given in the CNR system.

7.13 SUMMARY

Since all of the above systems vary approximately as the logarithm to the base 10 of the energy and all incorporate broadly similar day-night weightings, the following approximations can be used for comparison of the systems.

$$L_{dn} \simeq \text{CNEL} \simeq \text{NEF} + 35 \simeq \text{CNR} - 35 \simeq \text{NNI} + 25 \simeq \text{dBA}_{max} - 34 \quad (7\text{–}12)$$

7.14 ADDITIONAL NOTATION

The existence of numerous measurement systems and the need for a greater refinement has led to further descriptive terms, usually related to the A scale, but equally applicable to other systems. This consists of the technique of using a subscript number such as L_{90}, L_{50}, or L_{10} to describe the percentage of time when the given decibel level is exceeded, that is, 90, 50, or 10 percent of the time. Thus an L_{90} of 40 dBA indicates that 90 percent of the measured noise values fall above 40 dBA. L_{90} is often considered as the average background level. Similarly, an L_{10} of 90 dBA indicates that only 10 percent of the levels are above 90 dBA.

PROBLEMS

For all of the problems that follow, use the octave-band spectrum shown below the problem set.

7.1. Find the overall sound pressure level (refer to Chapter 1).

7.2. Determine the total loudness (sones) and the total loudness level (phons) for this source.

7.3. Determine the noisiness (NOYS) and the perceived noise level (PNdB) for this source.

7.4. Determine the dBA level of this source by two different methods.

7.5. Compare the answers to the above questions and explain any apparent anomalies.

7.6. An airport has approximately 200 landing and takeoff operations per 24-hr day. Fifty of these operations are at night. It is proposed that a curfew be imposed so that no flights will occur between 10 PM and 7AM. Estimate the difference in the NEF value if the 50 operations are (*a*) dropped altogether, or (*b*) shifted to daytime hours, assuming that all other conditions remain constant.

Octave Band Midfrequency (Hz)	Sound Pressure Level (dB)
31.5	70
63	73
125	78
250	80
500	85
1000	90
2000	95
4000	98
8000	95

Part II

NOISE CONTROL

8

Systematic Approach to Noise Control

8.1 THE ECONOMIC FACTOR

At this point in a typical student's progress through the study of engineering acoustics, he or she begins to wonder how all of the pieces fit together. Thus an attempt will be made in this chapter to assemble a systematic approach to the solution of a noise problem—to point out where to start, how to proceed, and the problem areas or pitfalls along the way.

It is very easy and very trite to say that all noise problems consist of three parts, that is, the source, the path, and the receiver. However, this approach is often taken in modern publications whether they be handbooks, textbooks, or periodicals. Such an oversimplification, in and of itself, fails to recognize a very important (and sometimes controlling) factor; that is, the economics of the situation. It is all very well to have a perfect engineering solution to a noise problem but if the cost of the remedy is far beyond the economic value of the product then the overall solution amounts to a mere academic exercise and will produce no results. In fact, nothing will be done. Conversely, if the noise level limits set by regulatory bodies are too low, the cost of reducing noise to meet those limits may cause defiance of the limits and in turn lead to lengthy and expensive litigation. And the high noise levels will still exist.

If the oversimplification of the problem into three parts is to be used, it should be modified in accordance with Figure 8.1, where the overriding

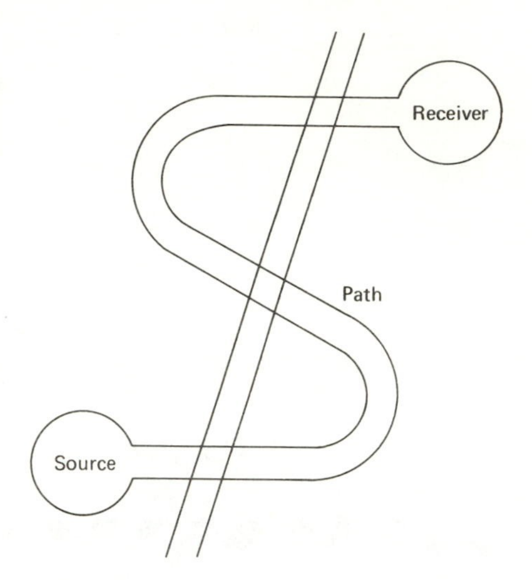

Figure 8.1 The noise control dollar ($) sign.

economics (dollar sign) becomes self evident. Cutting the path seems to be a popular way of solving the problem regardless of the cost factor.

This approach, however, tells us very little and is still unsatisfactory since it leaves many questions unanswered: How do I start? Where do I go? Do I quiet the source? Do I concentrate on the receiver? Do I move the receiver to a more distant location? and so on.

Even a simplistic approach to the problem would yield a schematic format as shown in Figure 8.2 where the step process is a little more detailed. The approach illustrated by this figure is typical of the thinking of many in engineering and associated fields. The format is still deficient in the economic factors and the user of such a format is led down the most direct vertical path to a preconceived solution such as "put it in a box." Little attention is paid to the possibly vital and cost-saving procedures of the side branches, particularly the one specified as redesign of product. The other side branch, ear protection, is in many cases is—or should be—the avenue of last resort, but it does have application where a technique of quieting the source is beyond the state of knowledge. For example, high impact sounds of very short duration and of a highly repetitive nature fall into that category at the present time. Research into this problem is currently going on, but interim protection of workers requires hearing protection.

In 1972 the Environmental Protection Agency of the federal government proposed a two-step schematic approach to the noise problem that is noteworthy. Among the purposes of that agency was to set future goals in its reply to an overall charge by the Congress to reduce environmental noise problems that exist in the United States.

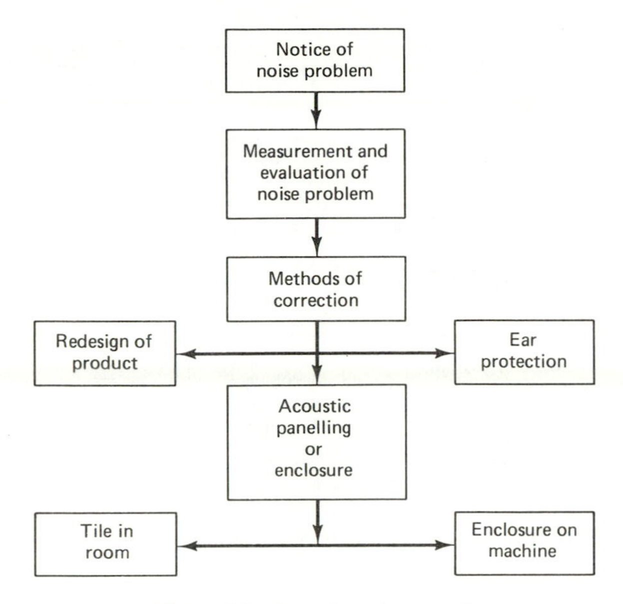

Figure 8.2 Steps for noise control.

The EPA presented the elements of a noise problem (Figure 8.3) in the standard format but itemized various portions of the diagram to indicate examples of each element. As can be seen, the economic issue was not addressed nor was there any direction given as to the areas where most of the effort should be directed to efficiently arrive at a solution. All solutions seemed to be contained by implication within the path element. Efforts toward a solution at the source were not indicated unless one assumed that the sources were listed in order of priority.

In an accompanying diagram, however, the economic issue was implied by the EPA. In this latter diagram, entitled "Elements of Prevention/Solution" and reproduced in Figure 8.4, the EPA indicated a two-

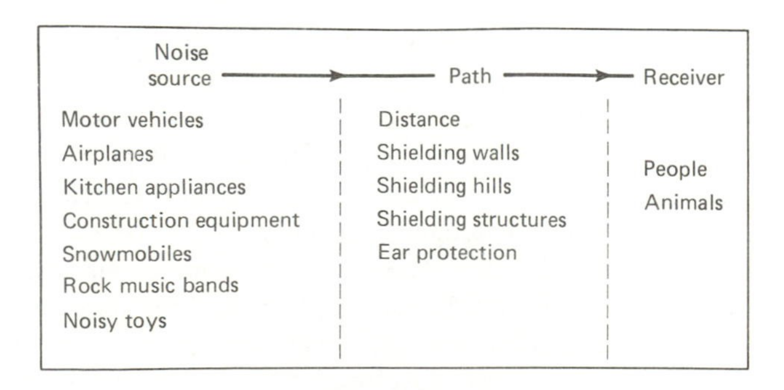

Figure 8.3 Elements of a noise problem. (Courtesy EPA)

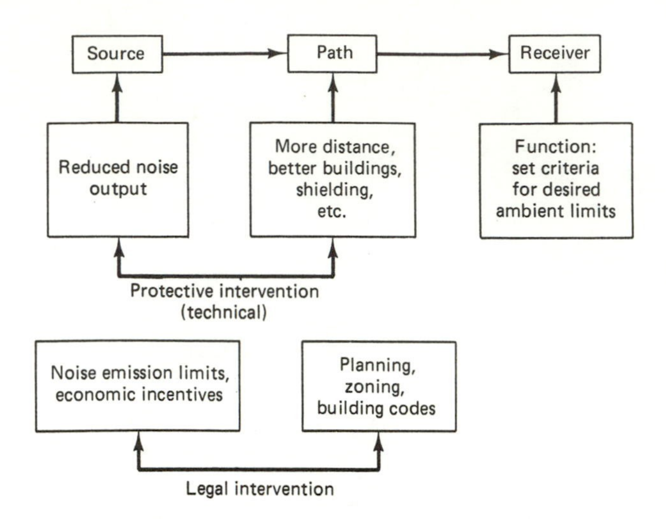

Figure 8.4 Elements of noise prevention and solution. (Courtesy EPA)

phase approach to noise control which serves today as a general guide to the activities of that agency. In this approach, both the engineering and legal efforts toward reducing noise are given equal attention. The economic incentives seem to be reserved to the legal profession when in fact these incentives should be included in the engineering efforts as well.

While all of these approaches are generally accurate, they do not go far enough in assisting in directing a systematic approach to an actual problem which will include economic factors as well as engineering design factors such as efficiency, long life, low maintenance, and so on. Thus, a further look at the approach to noise control is required.

8.2 THE SOURCE

Logic says that the starting point is with the source. Eliminating the problem at the source is the direct answer. Too often, it is assumed that the source cannot be reduced in noise level. Thus the path is used for a solution. The "put it in a box" solution is often used. This will accomplish the purpose of noise control *but,* while it may be the cheap method compared to other techniques, it is usually unacceptable to maintenance personnel. It also may destroy or reduce the life of the basic source unless air is artificially induced (creating more noise). In general, "boxing" the source should be the last resort rather than the first approach. It must be noted, however, that a great industry has been developed around the "put it in a box" philosophy.

We should observe at this time that we can always move the receiver away from the source. In industry, for example, providing automatic controls for a noisy machine will permit the operator to be positioned at a remote location. This does not, however, alleviate the basic problem—the source noise. Controlling the source is the first and usually the most effective step in a systematic approach to noise control.

A list of sources of noise is, of course, endless. It is possible, however, to categorize typical sources in different ways. One way is by type and their location in the environment. Each of these categories may require different measurement techniques as well as different control techniques. However, there are cases—with pneumatic tools, for example,—where the technique applied to noise control for those tools can also be applied to certain industrial machines. Note, however, that limitations on control may be exercised by federal, state, or local authorities, such as OSHA, FAA, planning, or zoning regulations.

A not all-inclusive classification of sources would appear as in Figure 8.5. In this form of classification, it can be seen that some sources can be found both indoors and outdoors and can appear in various environmental modes (see the solid line of Figure 8.5 for construction equipment) and some can appear in restricted modes (see the dashed line for industrial machines). From this illustration we can see that concentration on noise control of certain sources will be more productive and will have more of an impact on the receivers than will concentration on other sources. Such would be the case where a large industrial complex is to be analyzed and noise control procedures are to be applied.

Source noise can also be classified as to the type of sound emission

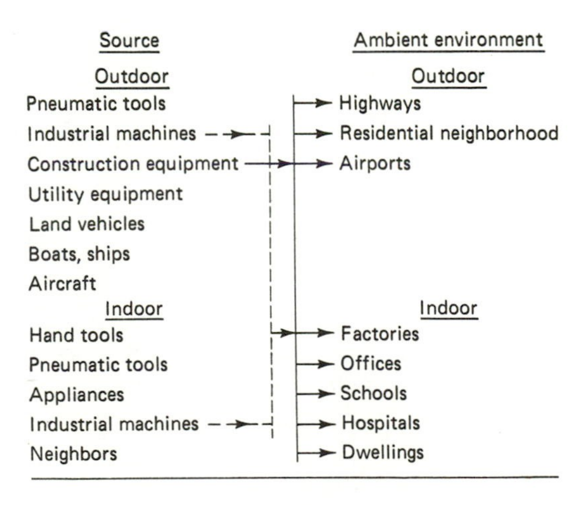

Figure 8.5 Classification by location.

STEADY NOISE

Without audible discrete tones	With audible discrete tones
Distant city noise	Circular saw
Waterfall	Transformer
Air-conditioning system	Jet engine compressor

NON-STEADY NOISE

Fluctuating	Intermittent	Impulsive	Impulsive quasi-steady
Heavy traffic	Aircraft flyover	Dropforge hammer	Rivetting
Pounding surf	Auto passing	Dog barking	Pneumatic hammer
Environmental—inner city	Train passing	Door slamming	Machine gun

Figure 8.6 Classification by noise characteristics.

that comes from a particular source, or, in other words, how it sounds to the receiver. Thus, there are those noise sources which can be classified as *steady noise,* (that is, where all frequencies are present and there is no fluctuation with time. These sources may be subclassified as to whether or not the sources contain audible discrete tones readily identifiable by the ear. Other sources can be classified as *non-steady noise* which would have subcategories such as: rapidly or slowly fluctuating sound pressure level; intermittent noises which occur in short bursts at random time intervals; impulsive noises which have a high level, short time duration (single gun shot), or impulsive sounds which have a high repetitive rate (automatic weapons) and thereby approach a quasi-steady noise condition. An example of this classification technique with typical noise sources is shown in Figure 8.6. The first step in the systematic approach to controlling a particular noise problem is to place that source in its proper category, both environmentally and by type of emission.

8.3 NOISE CONTROL

Refer now to Figure 8.7 where the noise control problem is broken down into three segments including the design concept, the design fix, and the acoustic fix. The latter two fixes will probably overlap in most cases and interact with each other while the design concept will set the constraints for the overall problem. The definition of the problem is usually general in nature and arises most often from a complaint of someone or some group segment of the population.

Identifying the sources does not usually require sound measurements at this juncture although those measurements will be required prior to attempting those procedures indicated in the "Design Fix" column. A reference to the use of Figure 8.6 is usually sufficient at this point.

Of great importance in the design concept stage is the correct identification of the population affected by the source, that is, the receivers and the path of propagation to them. Having identified the people affected by the noise source, it is then necessary to examine the acceptable noise limitations to be imposed upon or accepted by those individuals. Are these limits those which are established or imposed by existing criteria—either legal or otherwise—or are there peculiar conditions which makes this problem unique, for example, is there a conditioned population in a generally high noise environment such as that which exists in major urban centers.

Prior to a design fix and as part of the design concept, the restraints to the design must be examined. Are there legal limitations? Is the problem such that the maximum noise exposure for established time periods has been promulgated by such laws as the Federal Occupational Safety and Health Act (OSHA)? Is the maximum source level specified by a federal agency such as the aircraft noise emission limits established by the Federal Aviation Administration (FAA)?

The operational and maintenance requirements must be critically reviewed. Constraints imposed by required access to interior parts, say, to safety switches for either operation or maintenance, must be viewed at the design stage, and particularly if the design involves the attempted noise reduction of an existing piece of machinery. Consultation with experienced operators at this stage can save many potential redesign requirements after the design is fixed.

Design concept	Design fix	Acoustic fix
Define problem	Noise flow diagram	Structural
		Damping
Identify noise source	Effect on adjacent equipment	Vibration Isolation
Identify population affected		
	Satisfaction of restraints	
Identify acceptable noise limits		Airborne
Identify restraints		Reduce power
Legal		
Social	Limitations on product	Enclosure
Engineering		
Economic		
Identify alternate designs		Economic balance

Figure 8.7 Design procedures in a systematic approach to noise control.

We must address the social aspects of the problem. The esthetics of the final design should be considered here. Will the final design enhance the visual parameters of the problem? Will the result of the acoustic fix be a more appealing product? Giant noise control devices are often considered an ugly blot on the landscape. The decision to use such a device is often faced by engineers involved in the design of power plant facilities to be located close to their customers in inhabited areas. They must decide, for example, between the forced draft cooling tower with its large area but low height profile and the inherently quieter natural draft cooling system which requires a structure usually more than 500 ft high. The social implications of "no solution" should be examined as against a total or partial reduction of the source noise. To some extent the no-solution concept will have a bearing on the economic constraints imposed on the design concept.

Very early in the approach to a solution to a problem of controlling a noise source, the economic constraints must be spelled out. Noise control devices can often be very expensive. Low frequency attenuation is more expensive than the treatment of high frequency sounds. This cost must be weighed against benefits to be obtained. Noise control dollars are often considered nonproductive dollars at first glance by contrast with, for instance, heat recovery devices which not only save energy but pay for themselves in fuel costs. However, when these dollars are weighed against potential savings in compensatory awards for hearing loss or as the price of a good public relationship with neighbors, they take on a less frightening aspect. When a company with a severe noise problem which is hazardous to their employees is faced with a $500 per diem fine or a threat of injunction resulting in the closing of its operations, the cost of a noise control devices ranging up to $50,000 or more soon becomes economically viable.

When the noise control procedures are inserted at the design stage of a piece of equipment—or in the design of a building or highway—they can result in altered decisions for materials, processes, and so on, which can actually result in savings to the producer, owner, or general public as the case may be.

Case histories exist where shorter and more efficient air passages requiring less material (which are thus less costly) have been used. Different and cheaper bearings are all examples which could result in a quieter and at the same time less costly end product. It has been estimated that engineering or consulting costs to perform the above evaluations of design can add as little as one-tenth of one percent to the manufacturing or design costs. Thus to assert that all noise control procedures are costly is a distortion of the field of noise control.

8.4 ENGINEERING PARAMETERS

Having established all of the above constraints, it is now up to the engineer to examine the basic engineering constraints imposed upon the problem. *The basic purpose of the noise-producing device cannot be altered*. It still must do what it was intended to do.

If a machine is to be used for stamping metal parts, breaking up a pavement, drilling through rock, heading bolts or screws, or is to be a portion of a transportation mode, then these become basic engineering parameters. The engineer can, however, examine in detail the machine, process, and so on, and identify prime and subsources of the noise. He or she can at this point identify alternate methods for producing the same end result. These methods become alternate designs and not necessarily constraints. The engineer can examine mounting systems, radiation patterns, material use, and power transfer mechanisms. Each of these considerations opens new paths along which the resulting design concept can proceed. By now, the elements of the design fix begin to be established.

8.5 DATA ACQUISITION

Prior to proceeding with the first stage in the design fix (see Figure 8.7) it is necessary to obtain adequate sound measurements to achieve reliable data upon which to base design decisions. Without going into specific detail as to what instruments to use (this will be taken up later), it is possible to identify the necessary steps to obtaining adequate design data.

Figure 8.8 indicates a general set of steps for obtaining airborne sound measurements prior to proceeding with the design. In this figure, the measurements are classified according to proposed use. The sound measurements must include data to determine both the characteristics of the source and the sound levels that are representative of the surrounding or *ambient* noise field. Appropriate measurement positions must be chosen around the sound source to account for possible directivity characteristics of the source, that is, areas or directions of higher level sound propagation than others. At these positions, a complete spectrum of data—octave band at a minimum—as well as the measurement of the A-weighted scale (dBA) should be taken. At all measurement stations, data should be taken with the source both "on" and "off" to determine the extent of the source's intrusion into the general environmental background. The A-weighted sound measurements are used to determine compliance with design criteria requirements or legal constraints which must be met in the design fix.

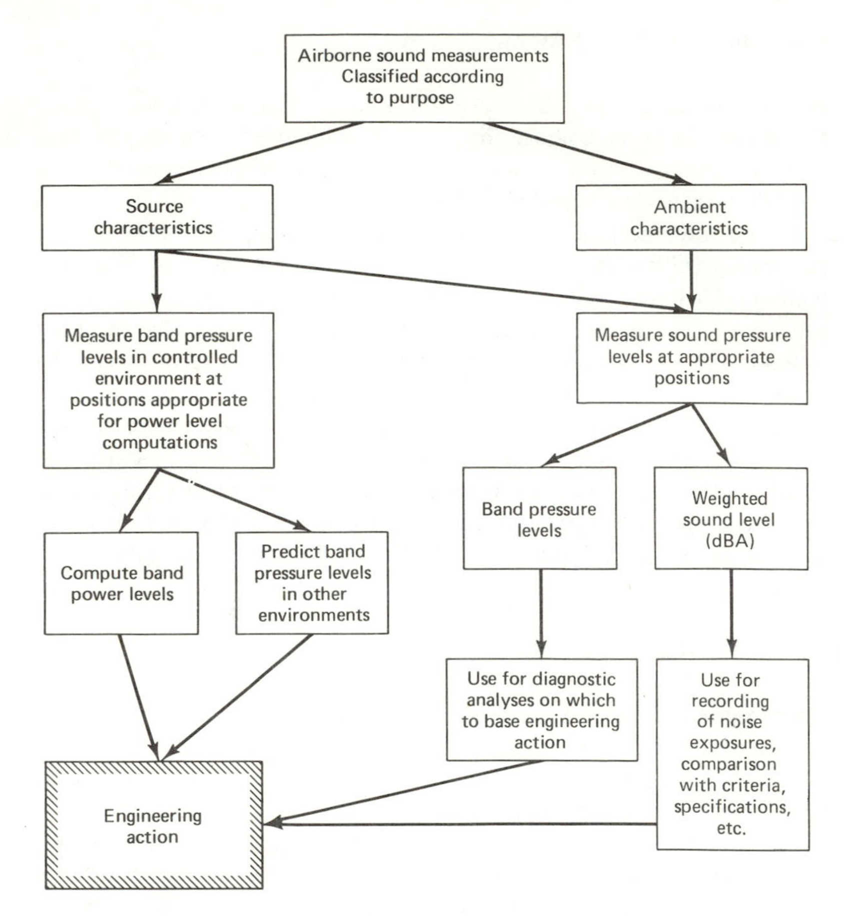

Figure 8.8 Steps for obtaining airborne sound measurements.

Band pressure levels of the source and any evidence of directivity are used in the analytical process upon which decisions are made as to providing for proper noise control devices. Prior to the determination of an engineering solution it is usually necessary that other environments be considered as potential locations of the source. Information about those environments must be available or obtained. Fans designed to be used in the home, for example, may be used in the bedrooms, the living area, or the kitchen. Each of these areas has a different ambient level but the fan must meet the most rigid criteria—that for the bedroom.

Figure 8.8 indicates the necessity for a multiple input of information (from all four channels) prior to adequate engineering action. While this statement seems self-evident, it is surprising how often engineering action is taken based upon insufficient input data. This is particularly true in many cases where solutions are proposed purely on the basis of only A-scale data without any evidence in the form of direct data regarding the spectral distribution of the sound energy. Such solutions are predestined to be in error and are not only poor engineering but many times result in great and unnecessary costs (economic waste).

Refer now to Figure 8.9, which illustrates methods for taking sound pressure measurements. This figure indicates the general approach and instrumentation required for each of the previously indicated measurements. Here the measurementation required for a specific purpose is noted as well as an indication of the location of the source and the type of environment required for reliable data. Generally survey measurements are made as the word implies to determine whether or not a basic noise problem exists. Field measurements imply that a problem does exist and that data are required to attempt an engineering solution. This also implies that the noise source is of the type that cannot be either moved into a laboratory controlled situation or be simulated in the laboratory. The measurements are therefore subject to weather conditions if outside or to other conflicting sources if inside. In an ideal situation, all noise sources should be measured under controlled laboratory conditions, but, of course, this is often impossible.

A data acquisition system which will serve all of the above purposes under all environment conditions as well as at all locations is shown schematically in Figure 8.10. Not all of the elements of the system are required for all measurements and other specialized equipment may be inserted at appropriate places for certain types of data acquisition. Discussion of the details of instrumentation will be given in the next chapter. However, we can see that the arrangement of equipment shown in Figure 8.10 can be viewed as layers. The top layer or the simple sound level

Measurement purpose	Instrumentation	Environment	Location
Survey	Sound survey meter	"As is" No control	Indoors or outdoors
Field	Portable instruments that meet requirements of applicable ANSI standards	"As is" or with minor changes	Indoors or outdoors
Precision	Laboratory instruments that meet USASI standards	Controlled -Anechoic- -Semi-anechoic- -Reverberant-	Indoors

Figure 8.9 Methods for sound pressure measurements.

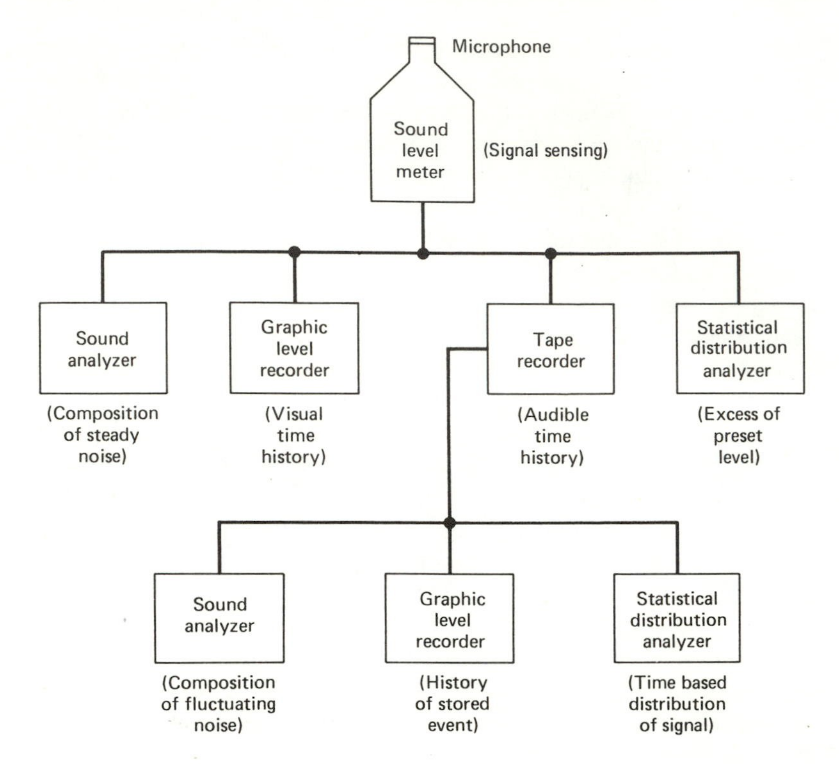

Figure 8.10 Data acquisition systems.

meter should be available to all people dealing with the noise problem, including salespeople and manufacturers of noise control equipment. The first and second layer equipment should be the minimum instrumentation available to the noise control engineer or consultant. The full complement is the minimum required for adequate research facilities.

8.6 NOISE-FLOW DIAGRAMS

After the design concept has been established and the sound measurements taken, proper engineering design for noise control can be initiated. Referring back to Figure 8.7, the engineer should now prepare an elementary *noise-flow diagram* with as much detail as has been disclosed by the sound data and by the configuration of the noise producing devices. The ability to accurately construct this type of diagram and the later implementation of it is one of the most powerful tools available to the noise control engineer. The purpose of the diagram is to identify those paths along which the sound can travel in getting from the source to air and thence to the receiver. It is often possible to quantify these paths or at least to rank order them in order of importance and thus achieve some orderly direction in the process of achieving the most efficient noise control design. Figures 8.11 through 8.14 illustrate different examples of flow diagrams. Sound may travel along or through the structure before being radiated to the air. Figure 8.11 illustrates the general concepts of a noise-flow diagram, and indicates that airborne sound may activate panels or other structural components through the resonance phenomenon. All paths, once identified, provide a direction for the noise control procedure. Ultimate control will depend upon damping or eliminating all paths. Remember that vibration isolation as well as vibration damping are key components in the search for a satisfactory solution to noise control.

Figure 8.12 shows a noise-flow diagram for a pneumatic chipper which is considered one of the noisiest tools in industry. The purpose of the chipper is to remove rough spots from castings. Noise is created when the blade comes in contact with the work piece. It is radiated from the point of contact of the blade as well as from the work piece (casting). It is conducted along the work piece and the tool itself and then radiated to the air. A secondary noise source is the exhaust port on the tool.

Figure 8.13 is typical of many product noise problems. The product being tumbled for removal of burrs or other imperfections becomes the source of the noise. Transmission occurs both through the air and through the structure.

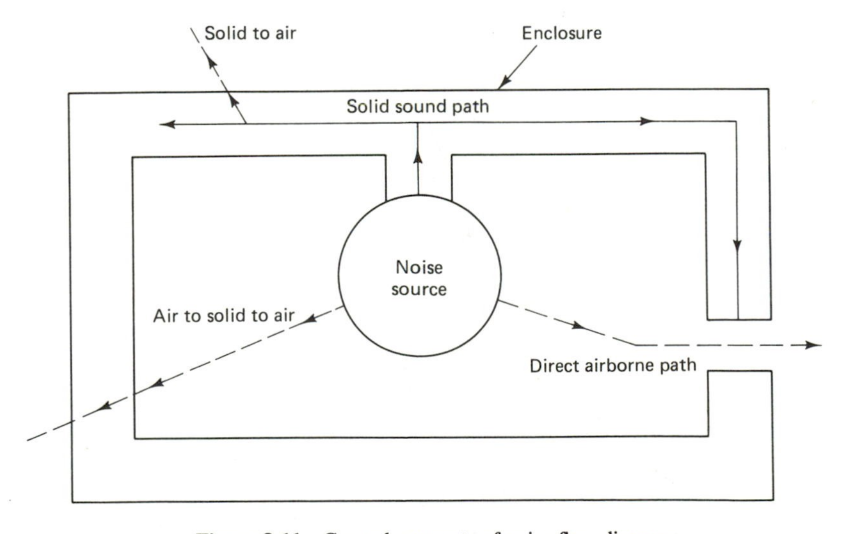

Figure 8.11 General concepts of noise flow diagrams.

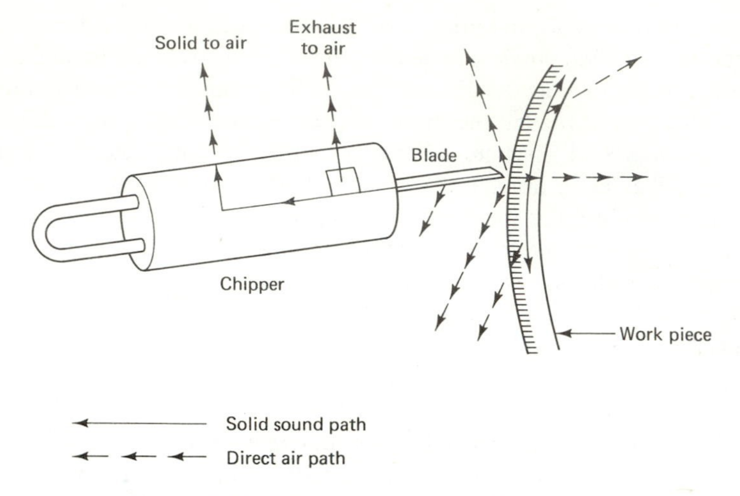

Figure 8.12 Pneumatic chipper noise flow diagram.

Figure 8.14 shows a schematic view of a simple electric motor. This is an example of a multiple source device. Each source can be traced to indicate its path to the receiver. Controlling one source (for example, the fan) will not necessarily produce a quiet motor.

Final action in the design stage is to reexamine the original restraints and the effects of the proposed solution on adjacent equipment or personnel

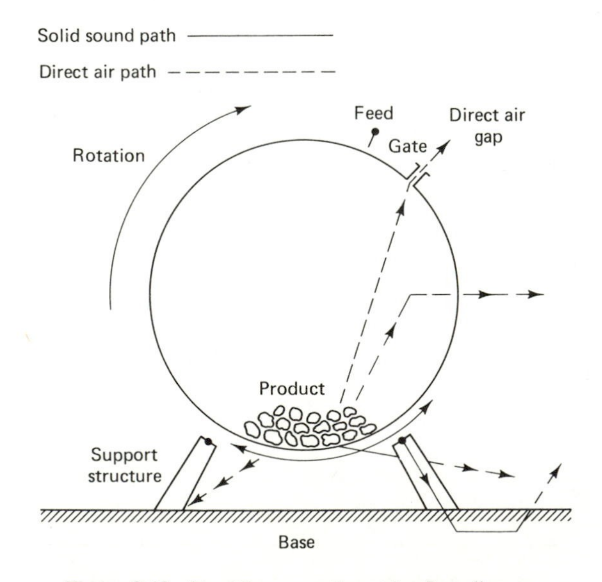

Figure 8.13 Tumbling operation noise flow diagram.

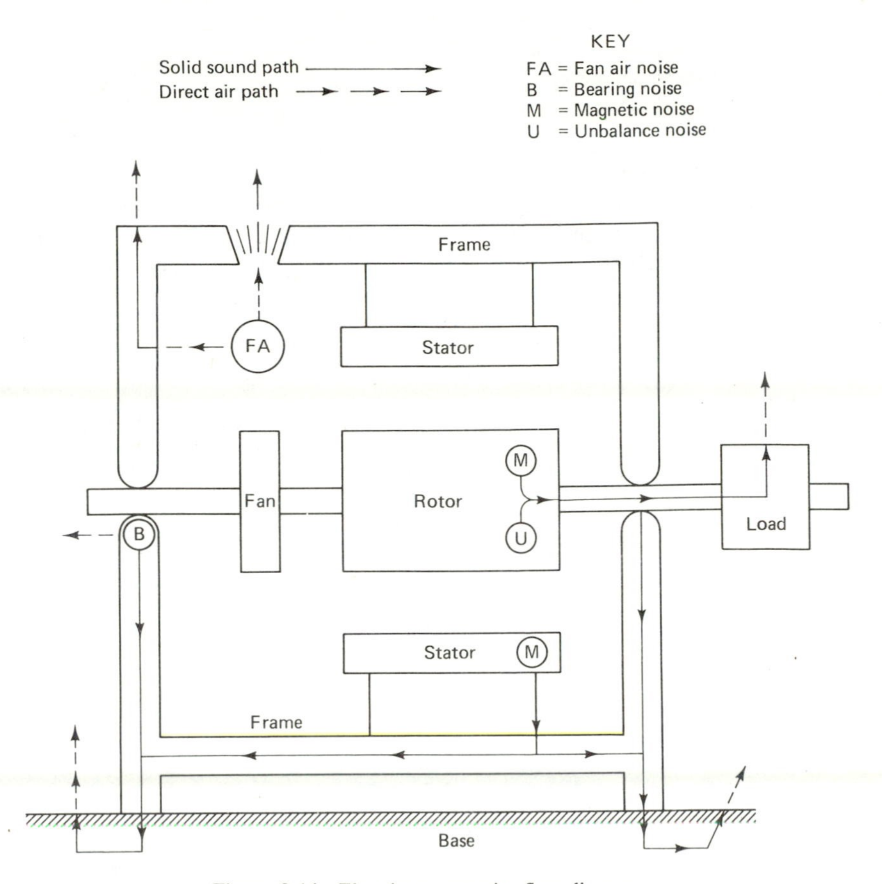

Figure 8.14 Electric motor noise flow diagram.

and to spell out in detail the limitations, if any, which the solution imposes on the original product. In particular, it is a vital necessity to balance costs of the control procedures against the economic benefits to be achieved if the control design is applied.

PROBLEMS

8.1 Classify the following noise sources in accordance with the headings shown in Figure 8.6.

- **a.** A powered lawn mower
- **b.** An electric golf cart
- **c.** A household food blender

d. Conversation
e. An automobile horn
f. A jack hammer

8.2 With respect to Figure 8.6, list at least one more source under each of the categories mentioned.

8.3 Choose some household appliance such as a food mixer, a food blender, or a sewing machine and construct a noise-flow diagram to be used in the analysis of the noise sources and paths within the appliance. Do not forget the product noise.

9

Instrumentation

9.1 INTRODUCTION

As we saw in the last chapter, the key to the proper understanding of a noise problem is the choice of proper instrumentation to measure and achieve a physical description of a particular noise source. As a corollary to the choice of instrumentation, the knowledge of the operation and the limitations of a specific instrument or instrument package is equally vital. While a great deal of specific information concerning all sound equipment is available in the literature published by the equipment manufacturers (and reference to this material is recommended), much of this material is expressed in the technical terminology peculiar to the electronics field. Much of it assumes that the person acquiring the equipment has a knowledge of the desired operational characteristics of a particular instrument and an ability to make judgments between pieces of equipment based upon this knowledge. This, in general, is not true and therefore often results in the acquisition of nonapplicable instrumentation for a specific purpose. Proof of this lack of knowledge exists in the ongoing proliferation of short courses sponsored by manufacturers teaching the fundamentals of sound propagation and control to educate the potential buyer. Product orientation, which is another way of saying sales pitch, is inevitable under these circumstances. This is not to downgrade this practice but to point out that, however sincere the manufacturer, the results of short courses often are misleading and can lead to a form of pseudo-expertise which can be harmful.

With the above thoughts at the forefront, the purpose of this chapter is to take an overview of the entire problem of instrumentation and to present limitations, guidelines, and suggestions which will be helpful in selecting the proper instrumentation for the measurement of a particular noise problem. Sufficient depth will be achieved to enable the novice to have a firm ground upon which to base decisions. The presentation of materials will begin with the heart of any sound measurement system: various types of microphones. Following this, special-purpose instrumentation will be discussed.

9.2 MICROPHONES—GENERAL

The most critical part of a sound measurement system is the microphone. The sensitivity of the microphone under varied environmental exposures such as temperature, humidity, and wind conditions will severely limit the accuracy of sound measurements. The type of construction used is a further limitation of the use of a particular microphone in specific applications.

Remember that all microphones are energy transducing devices. They convert one form of energy to another. They must be designed so that they will pick up and transmit faithfully sound energy in the form of a pressure wave with oscillations occurring at different rates (frequency) and in varying amounts (amplitude). The pressure wave must be transformed into an electrical signal which, as near as possible, reproduces the pressure wave characteristics with great fidelity. All of the electronic circuitry following the microphone is dependent upon this faithful reproduction of the original pressure wave into an electrical signal, that is, dependent upon the microphone's accurate performance as a transducer.

Attempts to develop efficient transducers for the accurate measurement of sound are believed to have their origin in the invention of the Rayleigh Disc in 1882. The Rayleigh Disc method observes optically the deflection of a freely suspended wafer exposed to a sound field. The particle velocity of sound is the measured quantity. Modern microphones respond to a difference in pressure across a diaphragm. This can be obtained either by a difference in the phase of the sound pressure on each side of the diaphragm, thus measuring the particle velocity of sound, or by maintaining constant pressure on one side of the diaphragm and hence measuring the sound pressure itself. The former type of microphone is known as a pressure-gradient microphone and the latter as a pressure-sensitive microphone. The pressure-sensitive microphone is the basis for the modern microphones used for sound measurement purposes.

Microphone type

Dynamic	●
Ceramic	O
Condenser	X

Properties	Best	Satisfactory	Least desirable
Stability	X	O ●	
Dynamic range	O X	●	
Sensitivity	●	X	O
Frequency response	X	O	●
Temperature	O X		●
Moisture	●	O	X
Magnetic	O X		●
Vibration	X	O ●	
Price	O	●	X

Figure 9.1 Comparison of microphone types.

Microphones can be classified into three general types depending upon their construction. The generic classifications are: dynamic or moving coil; ceramic, crystal, or piezoelectric; and condenser, which includes the *electret* microphone now coming into popular use. The characteristics of each type are compared in Figure 9.1.

9.3 DYNAMIC MICROPHONES

The action of a dynamic microphone has been crudely compared to the action of a mass suspended from a spring when it responds to an external forcing function. Oscillations occur in response to the stimulus and these are picked up and translated into electrical signals. Dynamic microphones are characterized by their rugged construction, making them suitable for the measurement of high intensity sounds and making them useful, or adaptable, in situations involving elevated temperatures. Water-cooled jackets have been used with this type of microphone to extend the temperature ranges in which they can operate.

The moving coil or electrodynamic microphone shown schematically in Figure 9.2 is one example of this type of microphone. The electrical output is the result of the motion of a conductor in a magnetic field. The

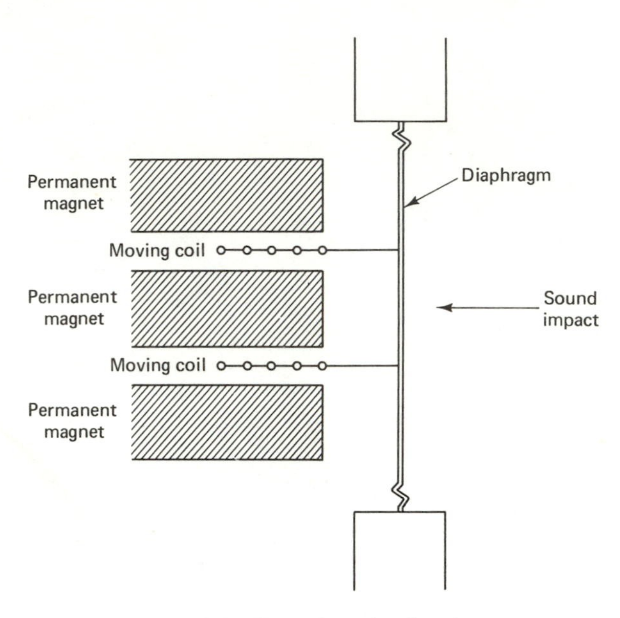

Figure 9.2 Electrodynamic microphone.

potential developed is due to electromagnetic induction and is proportional to the velocity of the conductor. For the velocity of the moving system to be independent of frequency, assuming constant sound pressure, the mechanical impedance must also be independent of frequency. This means that the impedance must be primarily resistive and not reactive. This is achieved by heavily damping the diaphragm so that the coil output is relatively smooth and does not have severe peaks and valleys.

Because of their inherent construction, dynamic microphones sacrifice frequency response and frequency linearity for the ability to achieve a high dynamic range with relatively good sensitivity. They are usually independent of moisture effects, but are highly susceptible to the influences of magnetic fields.

9.4 CERAMIC MICROPHONES

This classification of microphones contains those often referred to as crystal or piezoelectric. Many nonmetallic crystals become electrically polarized with the resulting deformation of the crystal shape which is the piezoelectric effect. Thus, if you visualize a crystalline cube, the pressure differentials set up across two opposite faces will create a difference in voltage (that is, voltage potential) across the two faces oriented at 90° to the faces

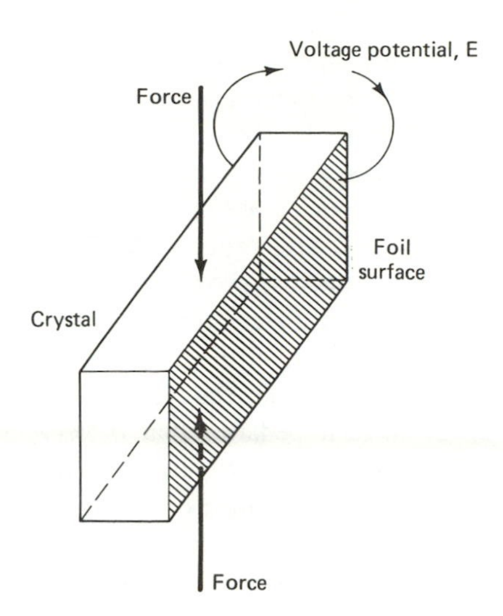

Figure 9.3 Piezoelectric effect (used in crystal microphones).

exposed to the pressure difference. This voltage will be proportional to the pressure difference. If a metal foil is cemented to these latter surfaces, the voltage potential can be applied to input circuits of an electronic amplifier as in the case of other electrical microphones. Figure 9.3 illustrates the concept of this type of microphone. The type of deformation of the crystal and its relation to the crystal axis and crystal planes determines the magnitude of the resulting potential. Deformation may be achieved by bending, shear, or compression. The alternating voltage appearing on the foil cemented to the surfaces is periodic in nature and is proportional (within limits) to the frequency and amplitude of the sound wave. Thus as the force reverses, the voltage potential will change sign and the current to the external circuit will reverse direction. Limitations on this type of microphone are associated totally with the particular crystalline material. For example, temperatures above 115°F will permanently injure a Rochelle salt crystal, one of the more common materials in use.

In general this type of microphone has the lowest price and, as can be seen in Figure 9.1, it has the best characteristics as an all-purpose microphone.

9.5 CONDENSER MICROPHONES

The condenser or capacitance microphone is one of the earliest precision acoustical instruments. Condenser microphones respond to a difference in pressure across a diaphragm. This difference can be obtained either as a difference in phase between the pressure at two sides of the diaphragm, as in the pressure-gradient microphones, or by mounting the diaphragm as one of the walls of a closed cavity, as in pressure-sensitive microphones. The nondirectional microphones used for sound measurement purposes are usually of the latter type. Figure 9.4 shows the principle of construction of the condenser microphone. The microphone cartridge consists essentially of a thin metallic diaphragm in close proximity to a rigid backplate. These two elements are electrically insulated from each other and form the electrodes of a capacitor. The air behind the diaphragm is in communication with the outside atmosphere only by means of a static pressure equalization hole, which has an acoustic impedance that is very large at audio frequencies. Variations in pressure due to sound waves will move the diaphragm which will vary the capacitance of the condenser at a frequency equal to that of the sound waves. A polarizing voltage is applied across the plates of the capacitor to induce a charge which remains constant because the time constant of the polarizing circuit is long in comparison with the period of sound pressure variation. Consequently an alternating voltage is generated in the condenser and by careful design it is possible to maintain the AC output voltage proportional to sound pressure over a wide range of frequencies and a large dynamic sound pressure range.

The sensitivity of the condenser microphone is inversely proportional to the total capacitance of the microphone circuit. Hence the total capacitance should not exceed the basic capacitance of the microphone cartridge by more than is absolutely necessary. For this principal reason, the con-

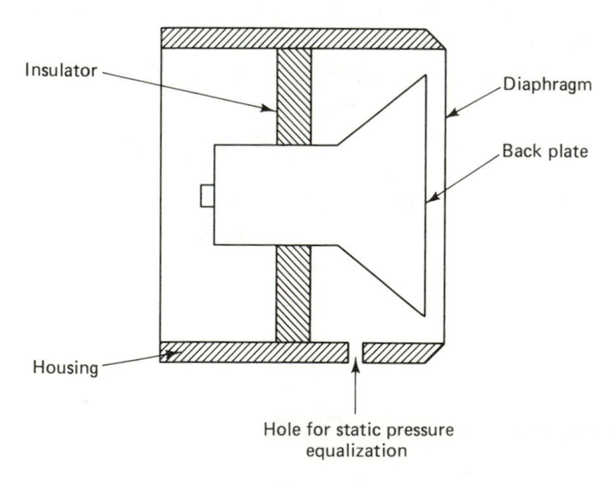

Figure 9.4 Condenser microphone.

denser microphone cartridge is normally mounted on the body of its preamplifier so that stray capacitances are minimized. A further advantage of mounting the microphone directly on its preamplifier is that noise induced in the cable transmission of the signal is eliminated.

In a practical microphone, the output voltage appears to be directly proportional to the polarization voltage. However, the gap between the diaphragm and the backplate is dependent on electrostatic attraction forces, so that variations of sensitivity with polarization voltage can occur. In practice, a highly stabilized polarization voltage on the order of 200 V DC is used so that the variation in sensitivity is virtually eliminated.

The size of the microphone has a bearing on its sensitivity. As the microphone diameter is made smaller, the sensitivity of most microphones tends to be reduced. There are two reasons for this: First, since the tension of the diaphragm is maintained independent of size, the reduction of the diameter increases the effective stiffness of the diaphragm; and second, the sensitivity also reduces because the variation of a capacitance is directly proportional to the diaphragm area. For measurements at higher frequencies and when more omnidirectional response is required the smaller microphone becomes necessary even though some loss in sensitivity must be accepted. A typical loss of about 13 dB in sensitivity will occur when going from a 1-in. diameter microphone to a ½-in. diameter microphone, although the frequency range will be increased.

Humidity effects create major problems with condenser microphones. An important reason for providing a release path from the inside of the microphone, aside from the obvious problem of excessive internal pressure such as can occur as altitude is increased, is to allow escape of water vapor, which can cause noise problems if condensation should occur. The influence of humidity is greatest when the microphone is moved from a warm, humid room to a cold environment. The diaphragm will obviously be the quickest part to cool because of its low mass and the water vapor inside the microphone volume will be rapidly cooled below its dew point. Since some heat will be provided from the more massive backplate, the most critical area will be at the unshielded edges of the diaphragm, and condensation here will cause partial or complete breakdown of the condenser dielectric, which will then appear as noise on the microphone signal. The change from a cold to a warm environment is not so severe in effect, since the warm moist air must first diffuse into the microphone over the period of time in which the microphone itself is gaining temperature. In this case, the condensation will occur first between the backplate and the diaphragm which will be the coolest part. Microphones can be equipped with adapters which will minimize this problem.

9.6 ELECTRET MICROPHONES (CONDENSER)

An electret microphone is a capacitor or condenser microphone using precisely the same basic principle to convert sound pressure into a varying electrical voltage. Because of the current interest and the future potential of the electret we discuss it separately from the general classification of condenser microphones.

An electret microphone is analogous to a permanent magnet in that it is an insulator that has acquired an electrostatic polarization. The existence of this effect was first recognized by Heaviside in the nineteenth century but the first use of the effect in a microphone is attributed to the Japanese physicist, Eguchi, in 1919. The life and stability of the first electret microphones was very poor because the prime dielectric was solid wax which, while easily charged, does not retain the charge.

Development work begun by Olson in 1929 and by the Bell Telephone Laboratory scientists as well as work in Japan and Great Britain, particularly in the field of materials, has led to the present use of plastic foils as a dielectric.

A dielectric is a material of low electrical conductivity and can therefore be classified as an insulator. In an insulator there are practically no free electrons to create a current when a voltage is applied. Instead the internal molecular structure becomes stressed and distorted. Thus a temporary polarization of the insulator occurs to produce a net positive charge on one side and a net negative charge on the other. Removal of the applied voltage, however, results in the molecules assuming their former state and the net charge disappears.

It is possible to arrange for a dielectric to retain a net charge for a period of time. The polarization caused by the application of an external voltage to a dielectric can be retained more or less permanently if the dielectric is made of a thermoplastic. This material is heated to a plastic state during the forming process at which time a high voltage is applied and maintained during the cooling process. After cooling, the voltage is removed and the plastic is left with a permanent charge, practically frozen into place during the process. The resultant dielectric is now an *electret*. See Figure 9.5.

Electrets eliminate the need for the polarizing voltage across the capacitor plates as in the common condenser. This results in a net reduction in the amount of electronic circuitry required. The backplate, which can be of varying designs, provides rigidity and reduces the possibility of the collapse of the diaphragm. Electrets are readily designed to withstand severe mechanical shock and internal noise. Humidity problems are reduced to a minimum. Only those conditions which can affect the foil itself can cause a problem.

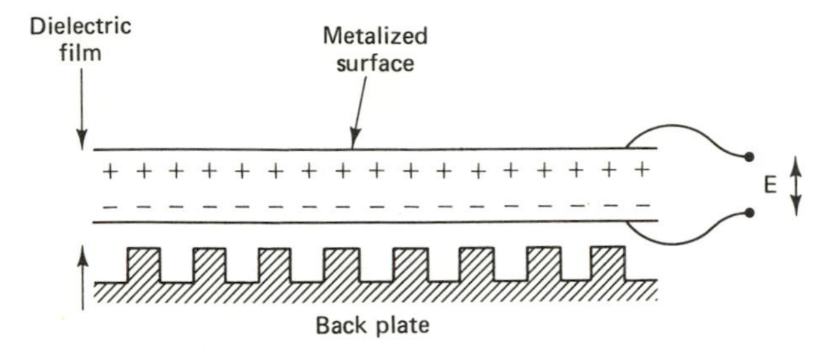

Figure 9.5 Schematic of an electret microphone.

The principal advantages of electret microphones are summed up as follows: relatively low cost, reliable, excellent transient response, low handling noise, and they can be made in very small sizes.

9.7 SOUND LEVEL METERS

An instrument designed to measure sound pressure called the sound level meter provides the simplest physical measure of the total amount of sound present. Such a measurement does not give an indication of the frequency distribution present in the sound nor any information on the human perception of the sound. But by relatively simple means it is possible to give this sound measuring instrument certain characteristics which make the results more understandable. Frequency weighting networks are introduced for this purpose.

The basic elements of a noise or sound measuring system consist of a microphone, a special amplifier or frequency analyzer, and a read-out or monitoring unit. The instrument may take many forms. It may be a small compact handheld instrument including a microphone and weighting networks, or it may be a laboratory-type stationary unit capable of automatically recording measured data.

The instrument generally referred to as the sound level meter is of the handheld type, and is supplied with a set of frequency weighting networks whose characteristics have been labeled *A*, *B*, and *C*. Some instruments also have a *D* network. The characteristics of each weighting network are shown in Figure 9.6. The *C* network has only a little dependence upon frequency while the *A* network has a pronounced frequency dependence below 1000 Hz. This latter network was previously discussed in Section 7.6. The characteristics of each of these networks have been set by international standards. By further international agreement, measurements made with any of the standard networks express the sound level in decibels (dB) with a further additional letter which denotes the scale

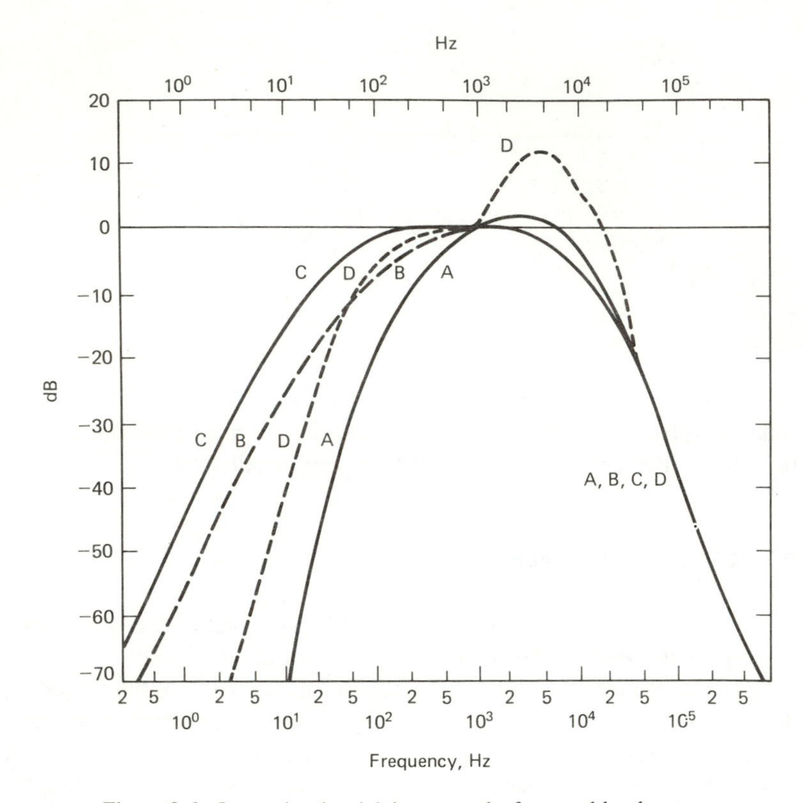

Figure 9.6 International weighting networks for sound level meters.

(or network) being used. Thus measurements on the *C* scale are recorded as dBC and those on the *A* scale as dBA. Measurements taken on all three scales can give some useful information regarding the frequency distribution of the sound being measured. If measurements on all three scales are approximately the same in decibels, then the frequency distribution is likely to be uniform. If there is a wide discrepancy between the *C* and the *A* scales (say 10 dB) then there is likely to be a heavy concentration of the sound in the low frequency range. This simple technique will serve to guide the observer as to the type of treatment necessary to control the sound. This is a technique that can be used for all sound level meters except those special-purpose meters such as in traffic monitoring devices which only present the *A*-scale levels.

The built-in meter read-out consists of an RMS rectifier and a meter circuit, the dynamic characteristics of which are as specified in the international standard. Two different meter damping characteristics, termed "Fast" and "Slow," respectively, are included in the instrument. The "Slow" characteristic is intended for use in situations where the reading taken on the "Fast" scale fluctuates too much (more than 4 dB) to give a reasonably well-defined value of the desired sound level.

The meter is equipped with a decade attenuator setting which establishes the value of the zero point on the face of the meter, that is, 60, 70, or 80 dB as the case may be. The meter face is scaled logarithmically (in a few cases linearly) to cover the intervening steps in any decade. Thus the reading from the meter face is added to the decade reading to give the final output. All decibel readings are referenced to a pressure of 0.00002 N/m^2. Some meters are now being equipped with a digital readout to facilitate quick readings with less possibility of error.

9.8 FILTERS—OCTAVE (1/1) BAND AND ONE-THIRD (1/3) OCTAVE BAND

Instrumentation necessary to perform a frequency analysis of the sound source consists of the use of filters of varying types. All filters in use can be characterized as band-pass filters in that only certain frequency ranges are passed into the measurement circuit while all other frequencies are rejected.

The frequency analyzers currently in use are of two types: Constant percentage bandwidth analyzers are those where the width of the analyzing filter (that is, the filter bandwidth) is a constant percentage of the band center frequency, independent of the absolute value of that frequency. Constant bandwidth analyzers on the other hand have a filter bandwidth that is a constant number of frequencies (Hz) wide. The bandwidth is not proportional to the center frequency. Normally the bandwidth of constant bandwidth analyzers is very narrow and allows the determination of a frequency spectrum in great detail. Such detailed analysis is important where harmonic distortion in electrical or electroacoustical equipment is to be investigated; when analysis is made of the vibration response of randomly excited mechanical constructions containing a number of very lightly damped resonances; or when a search is required to locate predominant peaks in the sound spectrum in order to correlate them to mechanical actions in complex machinery.

In the case of acoustic noise measurements, too-detailed information on the noise spectrum is normally not desired or economically justifiable. Thus the constant percentage type frequency analyzers are usually preferred. There are two basic reasons for this. Slight instabilities in noise-producing machinery mostly produce a constant percentage type of frequency distribution rather than a constant absolute type, and the measurement errors introduced by such instabilities are basically the same over the complete frequency range of measurement. Second, the ear responds to sound in much the same way as a constant percentage bandwidth analyzer having a bandwidth of about one-third octave.

Following the above train of thought, the one-third octave frequency analyzer should be the most suitable type for acoustic noise measurements. But this is not always true. Some noise criteria (such as community noise criteria) are based on one-octave (1/1) band data. These data are normally much easier to take and less time consuming (and therefore more economical) to obtain than the one-third octave band data when measurements are made in the field. On the other hand, when measurements are made of the noise produced by a specific machine, with the intent to investigate what changes can be made in the design of the machine to most effectively reduce the noise, one-third octave band data is essential. In fact, even narrower frequency bands may be required.

The kind of noise being investigated influences the choice of analyzer. If the noise is of the type producing continuous frequency spectra (community noise, air-conditioning noise, and even jet engine noise) which varies only little in level with frequency, then one-octave band analysis is completely adequate to describe the noise. However, noise spectra containing more or less sharp peaks should preferably be described in terms of at least a one-third octave band analysis. The determination of loudness should be made with a one-third octave band analyzer, although approximation techniques using the octave band method or the use of the A scale are available.

The choice of filter or subsequent read-out depends to a certain extent on the particular noise measurement problem at hand. A number of questions must be considered. What type of noise is involved? What time is available for the measurement? What is the aim of the investigation? What measurement accuracy is desired? When the noise is more or less of the steady type, and the aim of the investigation is to obtain an estimate of the noise level, the read-out provided by the instrument meter of a precision sound level meter or an octave band analyzer, using its "Fast" or "Slow" characteristics may be quite satisfactory. If, however, the noise is transient or intermittent (flyover aircraft or passing motor traffic) the use of tape recording or graphic level recording or statistical analysis would be the appropriate read-out. Transient, intermittent, or fluctuating noises may be recorded by experienced operators using the A scale of the precision sound level meter but some degree of accuracy is sacrificed.

9.9 AUXILIARY INSTRUMENTATION

Several pieces of auxiliary instrumentation have been developed and are being introduced regularly to perform special tasks designed to make the noise analysis more complete and accurate. The entire list is too lengthy to be discussed in this chapter; however, three very important pieces of equipment will be noted at this time. Greater depth of descriptive material is contained in literature published by the various manufacturers.

Magnetic tape recorders have become one of the most useful and powerful tools for the engineer involved in noise analysis. Data from remote locations can be recorded and transmitted to the laboratory for detailed analysis. As long as the tape is carefully calibrated—both for sound pressure level and frequency—the data can be reproduced in many different forms, that is, graphic printout, digital printout, real-time analysis, and statistical distribution. The tape recorder provides a means for storing and reproducing transient events. It can record time-varying noises as well as either short bursts or long periods of sound exposure. Sounds on tape can be deliberately distorted by altering the playback speed or by filtering, clipping, or adding noise. At times this can be useful in the process of noise analysis. Subjective response to similar sounds can be aided by having the sound on reproducible magnetic tape. In short, a carefully made tape recording on a high fidelity recorder can reproduce and store any sounds for later analysis or review.

Graphic level recorders are nothing more than a specialized form of *X-Y* plotter. Most are servo-type strip chart, pen recorders, having a wide range of speeds for the chart drive, a wide range of averaging times for the detector system, and a wide range of levels for recording. The continuous strip output has in its linear dimension a time base set by a combination of the paper and writing speed. The width of the paper establishes the range of decibel readings being continuously recorded. In another mode of operation, the linear dimension can be set to respond to frequency distribution while the width establishes the amplitude of the sound level at a particular frequency. Thus a frequency spectrum can be generated. In all graphic level recorders, the amplitude of the fluctuations is controlled by the frequency averaging time and the absolute bandwidth of the input signal. Graphic level recorders may receive signals directly from the output of a sound level meter (filtered or not), from a tape recorder output, or from the output of a real-time analyzer. In some cases the recorder can in turn be used to drive a statistical distribution analyzer to establish the distribution (or histogram) of the various levels of sound in a given time interval. Detailed operation of each graphic level recorder will vary with the manufacturer and users should become aware of these details through references to the manufacturers' literature.

Real-time analyzers provide another method of frequency analysis. Real-time or parallel analysis has become a powerful tool in the field of noise analysis. Here the output from a number of parallel filters is fed simultaneously (or within fractions of a second) to a read-out display such that the complete frequency spectrum for the range of the filters can be presented instantaneously. Very rapid changes in the spectrum with time can be readily detected. The display appears on a television-type picture tube which can be read visually, photographed, or stored through a series of outputs such as an *X-Y* recorder, graphic level recorder, or digital data receivers such as punched tape, printers or on-line computers. Connecting the real-time analyzer to a suitably programmed computer forms a flexible and very rapid system for automatic data processing with the advantages of real-time analysis.

The idea of parallel analysis or, generally speaking, simultaneous measurements in several measurement channels may not only be utilized for noise investigation purposes but it is also ideally suited for noise monitoring purposes. The only drawback is the bulk and cost of the system and therefore while it has been used for such, it makes an expensive airport monitoring system. Costs can be reduced without loss in accuracy by tape recording field data and using the output of the tape recorder to activate the real-time analysis system.

Since many sources of supply exist for real-time analysis systems, the user is again referred to manufacturers' literature for specific operational details on a given system.

Narrow-band spectrum or Fast Fourier Transform (FFT) analyzers are the newest analytical tool available for a detailed analysis of a sound (or vibration) frequency spectrum. These analyzers are related to but are an improvement of the real-time analyzer discussed previously. The real-time capability is maintained. The constant frequency bandwidth has been modified from a one-third octave sampling of 33 frequency channels to some 400 channels over the same or a slightly expanded frequency range. Most analyzers have a "zoom" feature in that selectable frequency ranges, say from 0 to 10 Hz to 0 to 20 kHz, are available. Within those ranges the roughly 400 channels are available. Within these channels a sampling rate of over 1000 bits of information concerning the input signal is available, producing an effectively instantaneous, real-time spectrum on the output display.

Many sources of supply exist for this type of analyzer. Each manufacturer specifies one or more features that a competitor lacks. Thus, it is again advisable to consult their literature for specific details.

PROBLEMS

9.1 Choose the necessary instrumentation to measure and record the reverberation time of an auditorium.

9.2 What type of instrumentation would yield information which would determine whether or not sounds from the stage are heard uniformly throughout the audience in a concert hall?

9.3 What type or types of equipment would be used to measure the soundpower output of a hydraulic pump?

9.4 While monitoring the noise from a fluctuating source of noise, you suddenly detect a crackling noise in your headset and observe a wildly swinging needle on your meter. What is the probable cause? (*Hint:* You are using a condenser microphone.)

9.5 What elements of a set of noise-measuring instrumentation would be used to measure (*a*) traffic noise, (*b*) a waterfall, and (*c*) a circular saw?

10

Community Noise

10.1 INTRODUCTION

The community noise problem is probably the most complex, pervasive noise problem affecting the largest population at the present time. The problem is international in scope and interest. Community noise is concerned with noise levels that are well below those considered as incipiently damaging to hearing for long exposures (that is, the OSHA limit of 90 dBA). The levels are those which can cause irritation, possibly speech interference, and loss of sleep.

The problem of outdoor noise is a complex mixture of highway and aircraft noise, of noise from factories, construction noise, and people noise. Noise sources which are stationary as well as those which are moving are involved. Variations exist in the types of noises, the level of intensity of the noise, its frequency characteristics, the time of day or night of the occurrence, and, equally important, the variation in the susceptability level of the various segments of the population to the noise exposure. In order to cope with community noise, the acoustical practitioner must draw on all phases of the science as well as a wide range of measurement techniques and must deal as well with the psychological and sociological aspects of human reaction to noise both as individuals and by population elements or components such as rural, suburban, and urban divisions.

A key element in the control of community noise lies in the development of adequate zoning for land use. This type of zoning involving noise lies under the general heading of performance zoning where specific maximum limits are established for particular land uses. Prior to setting these limits, however, it is mandatory (for the successful application of the zoning regulations) that detailed knowledge of the existing noise parameters—levels, reactions, and so on—be obtained. This is not a unique requirement since many aspects of performance zoning regulations require prior data before the promulgation of regulations (for example, wetland zones are only defined after field or geodetic surveys). Thus the first step is to provide for community noise surveys of sufficient depth and accuracy to provide a detailed knowledge of the many variables as they exist in real time. A simple straightforward measurement and analysis technique is not readily available. The technique is necessarily made complex by the multiple reactions and cross relations that exist in the basic problem.

10.2 HISTORICAL PERSPECTIVE

Prior to a detailed discussion of the problem of community noise, we will review the history of the attempts to solve this problem in the United States, Great Britain, Germany, Switzerland, and Japan. Such a review in detail would constitute a lengthy treatise in and of itself. Thus this perspective will only touch on the highlights.

The first recorded attempts to measure and evaluate community noise were the audiometer surveys of New York City noise on the streets and in skyscrapers as reported in the work of E.E. Free in *Forum Magazine* in New York City in the years from 1924 to 1928. After this, in 1929, New York established a Noise Abatement Commission which used crude (by modern standards) instrumentation to perform evaluations of the noise and make recommendations. New York's work in this area is continuing to this day and includes many attempts by citizen groups such as Citizens for a Quieter City which sponsored a detailed, block by block, noise and correlated sociological survey of some 24 city blocks on the West Side of Manhattan from 1970 to 1972. The raising of citizen consciousness by this technique has led to a more rigid enforcement of New York's present regulations.

In Chicago in 1974, after many previous surveys, the Department of Environmental Control began a program to determine the baseline noise levels within the city. Their procedure provides a continuous 24-hr signature in dBA at preselected locations with the data capable of being expanded to depict a statistical plot of noise for every 15 minutes of the

24-hr period at every location. Reference to this study will be made later in this chapter.

Another attempt at gaining knowledge of existing noise levels was the study by the Building Research Station, the London County Council, and the Central Office of Information of 36 square miles of central London, England, in 1963. Here, noise measurements were made at 540 equally spaced points, 1500 ft apart, and correlating data on the reactions of residents near those points were obtained in a social survey. This is the first recorded survey which departed from previous surveys where sound level readings were obtained individually from sound level meters. In this case the data was automatically sampled around the clock by use of tape recorders. The data was then analyzed and presented in the form of dBA readings. The "Wilson Report" of 1964 attempted with some success to translate this survey into recommended levels for various types of zones ranging from hospital zones to zones adjacent to major traffic arteries.

In Germany in 1938, initial techniques for developing a *noise map* where contours of equal loudness were drawn from field measurements were undertaken independently in the cities of Charlottenburg, Duesseldorf, and Dortmund. In the latter case, 1449 separate locations were studied to achieve the detailed loudness contours, an indication of the work involved to establish such baseline data.

By 1964, the Swiss had achieved sufficient data from physical measurements as well as social surveys to promulgate the tolerated outdoor noise levels for both day and night in at least two situations, that is, one for rural and urban areas away from main road traffic and industry and the second for urban areas near main roads and heavy locations.

In Tokyo since 1965, in each of the city's five zones (suburban, residential, commercial, light industry, and heavy industry) continuous surveys have been going on. In each of the zones, approximately 17 typical areas, each of about 25 acres, were selected and noises were measured in dBA at 20 points in each area. Readings were taken at least 50 times for intervals of 5 s to obtain an average and a range. Temporal distribution over 24 hrs was also determined. Correlations with traffic density and noise were obtained. Tokyo was one of the first cities to erect noise monitoring displays at critical points to indicate to the passersby the instantaneous level of that site as it was being observed. An unfortunate commentary can be made at this point in that despite these efforts, many visitors to Tokyo have rated it as a very noisy city and some experienced observers rate Tokyo as the noisiest city in the world. This is certainly a debatable point when one considers the noise levels of other metropolitan areas of the world.

10.3 PERFORMANCE ZONING—NOISE

The term *performance zoning* as applied to any subject implies the ability to establish zoning regulations on land use which are meaningful, capable of scientific interpretation, able to be applied uniformly by standard engineering techniques, and which have both an economic and sociological acceptability. Planning and implementation by zoning regulations are essentially protective measures. Everyone must be protected from noise interference with his or her privacy. The control of noise as a nuisance must be exercised wherever the source of the noise originates. Care must be taken not to overcontrol the noise through unnecessary or burdensome criteria. Zoning regulations must therefore take into consideration the source, the receiver, and the path between in a manner which will control and regulate the noise in an economic and equitable way.

Performance standards exist in many fields other than noise. Examples can be found dealing with structures, materials, and sanitation. Thus, to exercise control over noise is not a new concept. Present practices, however, have been too loose on the one hand, simply outlawing unnecessary noise, or too restrictive on the other hand by selecting sound level numbers arbitrarily and writing them into a zoning code. Either practice is equally undesirable. "Unnecessary noise" is too subjective and numbers may produce requirements which are in fact below existing ambient or background noise levels.

Most communities have ordinances declaring specific noises as a nuisance, and thereby offenses that are classified as disturbances of the peace. Very few communities have enforceable performance zoning codes regulating the continuous or invasion-of-privacy type of noises. Many towns that say anything at all, simply state that no industry shall make "unnecessary noise." They say nothing about other producers of noise, such as trucks, air-conditioning units, aircraft, commercial establishments, construction noise, and so on. No account is taken for day versus night operation, for frequency, or for measurement position.

The application of the concept of performance zoning at present is extremely diverse. Some towns, for example, have adopted specific noise level limits for their industrial parks. These are based upon determination of existing background levels and the projected type of development desired in their parks. The procedures account for daytime and nighttime operation. The applicable numbers for each town vary because of the peculiar locations and *the sound levels established for one town are not applicable to any other*. These towns, however, generally have not gone beyond the industrial park to apply their performance zoning. This is a weakness in each town's efforts. Such zoning might be challenged as discriminatory.

Similarly, towns have zoned against air-conditioning noise and no other, while other towns have adopted the same criteria for noise levels throughout the entire area of the town. Each of these approaches can be challenged as either discriminatory or impractical. Exemptions for motor vehicles, trains, or aircraft only serve to weaken the regulations.

Each of the above approaches is unnecessary since the equipment, technology, and trained personnel now exist to accurately measure such important parameters of noise such as Sound Pressure Level, the frequency spectrum, the rate of repetition, and the location of the source by scanning techniques. Sufficient criteria exist to compare a given local situation with that of other comparable areas. Thus, a given geographical area can be assigned specific maximum noise levels depending on its projected or actual use, which if exceeded will result in a 90 percent probability of complaint from the average individual. Engineering techniques and design data exist in sufficient quantity and with sufficient validity to allow for the application of specific criteria prior to construction or to the use of rental space in existing buildings. Currently, industry is in favor of, and in many areas is demanding, performance standards. Increased productivity and better public relations are both byproducts of good noise control. There is no further need to rely on the empirical "reasonable person" to act as a judge of noise nuisance, thus avoiding one sure step towards litigation.

In the event of complaints, available methods of sharing the costs have been worked out and specified in zoning codes. To simplify this thought at the risk of oversimplification, it can be stated as follows: After complaint on excessive noise arises, the planning and zoning board assigns a qualified consultant to investigate. If the complaint is valid, the noise maker pays for the investigation; if the complaint is invalid, the municipality pays and the fee is collected from the complainant by other means. This reduces the risk of unreasonable complaints.

Planning and zoning boards are ready to accept the concept that every area is unique and that while numerical standards exist, they must be discreetly applied and used primarily as guides. The boards are ready to recognize that many of their past problems, including legal litigation, have resulted from loose wording, from single number (dB, not dBA) ratings, and from no inclusion at all or from such phrases as "no disturbing noise." The boards are also ready to recognize that too much interpretation has been made by the laypeople, including attorneys, not trained in the science. They further recognize that there exists in almost every state of the union a well-trained and experienced body of consultants who are available at reasonable professional fees.

10.4 PREDICTABLE VARIABLES

Neighborhood response to community noise is a complex variable and only through the application of statistical methods can any success be achieved in predicting that response. The factors involved in the neighborhood noise are identifiable. Norms for one neighborhood will vary both within the neighborhood and with its adjacent population areas for sociological and psychological reasons. A previous history of noise exposure will be a major contributing factor to neighborhood response to new noises. Thus,

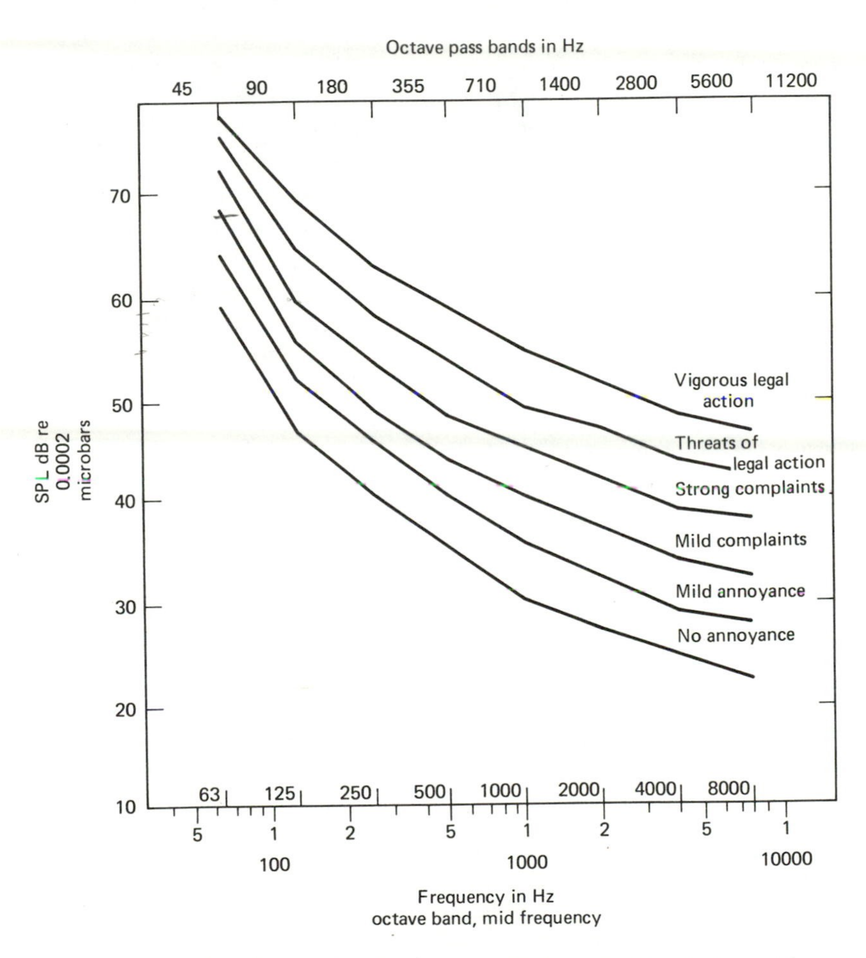

Figure 10.1 Response (at the 90 percent probability level) of a suburban neighborhood to continuous nighttime noise.

quiet rural areas will rise up in protest at the thought of a new highway or airport planned nearby while city dwellers might give little thought to such increases in noise. Additional factors governing the response will be: source sound power radiated at different frequencies; the presence of predominant single frequency components in the source; the presence of sharp impulses (gun shots, rivetting, drop forges, and so on); the time duration of the noise; the time schedule of the noise (day or night); and the repetition rate of the noise.

All of the above factors have to do with the concept of loudness as discussed in Chapter 7. The loudness of a noise is an attitude or state of mind and care must be taken to account for this factor in the response to community noise. Annoyance cannot be reliably predicted for any given individual at any given time. Probabilities of annoyance, however, can and have been predicted with success as judged by reduced complaint levels, wherever the probability concept has been applied.

Working with a 90 percent probability factor, an example of neighborhood reaction can be illustrated by use of Figure 10.1. In this figure, which is a plot of sound pressure levels as measured in octave frequency bands near residences, a series of *probability zones* are established. These probability zones represent the 90 percent probability point of response to continuous noise at *nighttime* in the *summer* for individuals living in a *suburban* neighborhood. Thus as the measured levels rise, the probability of action increases. To use this set of curves for *daytime* exposure, all values should be increased by 10 dB.

In this particular neighborhood, the sounds from the environment (designated at the ambient level) ranged within plus or minus 3 dB of the lowest curve. To apply this figure to continuous noise situations in other neighborhoods would require the establishment of the ambient level in the field and then setting the lowest curve to fit that data. All other curves would be maintained their same relative distance apart.

TABLE 10.1 Corrections to Figure 10.1 to Account for Variation in Neighborhoods, Nighttime Summer Criteria (for daytime use, all curves should be raised 10 dB)

Neighborhood	Correction (dB)
Very quiet suburban	Down 5
Suburban	No correction
Residential—urban	Up 5
Urban—near industry	Up 10
Light industry	Up 15
Heavy industry	Up 20

TABLE 10.2 Corrections to Figure 10.1 to Account for Variations in the Character of the Noise

Variations in Character of Noise	Correction (dB)
Winter (average)	Up 5
Continuous	No correction
Presence of pure tones	Down 5–10
Presence of impact or short duration bursts	Down 5–10
Repetitiveness of pure tone or impact	Down 5–15
Previous exposure (or conditioned exposure)	Up 5

10.5 NEIGHBORHOOD CLASSIFICATION

Neighborhoods may be classified as shown in Table 10.1 and in general the level of the bottom curve of Figure 10.1 can be adjusted up or down by the amounts indicated in the table in order to predict the response to be expected if additional noises are introduced to the neighborhood. It is well to repeat that Table 10.1 is for continuous noise only.

It is well known that other types of noises exist and therefore it is necessary to perform other corrections to Figure 10.1 to account for these variables. Table 10.2 serves as a guide to account for the common variables and deviations for continuous noise. Table 10.3 indicates the corrections to Figure 10.1 which should be applied to account for the time of exposure to the noise. It is interesting to note that the shorter the duration of the noise, the lower is the tolerance level to that particular type of sound. This is related to what has been named the *intrusion* factor of the noise. The greater the intrusion, the greater can be the expected community response.

TABLE 10.3 Corrections to Figure 10.1 To Account for the Time of Exposure to the Intrusive Noise

Time That the Source Operates in Any 8-hr Period (in min)	Correction (dB)
1	Down 20
10	Down 15
30	Down 10
60	Down 5

10.6 EXAMPLES OF COMMUNITY RESPONSE

Examples of the observed community response as reported in the literature for residential areas exposed to various types of noise sources are shown in Figure 10.2. In this figure, three typical case histories of the response to noise in residential areas are shown. The same key number is used to identify both the source level as measured in the residential area and the normal background or ambient level of that area. The three cases are described in the following paragraphs.

Case 1 concerned a large wind tunnel test facility for jet engines located in the midwestern United States. The response was so vigorous

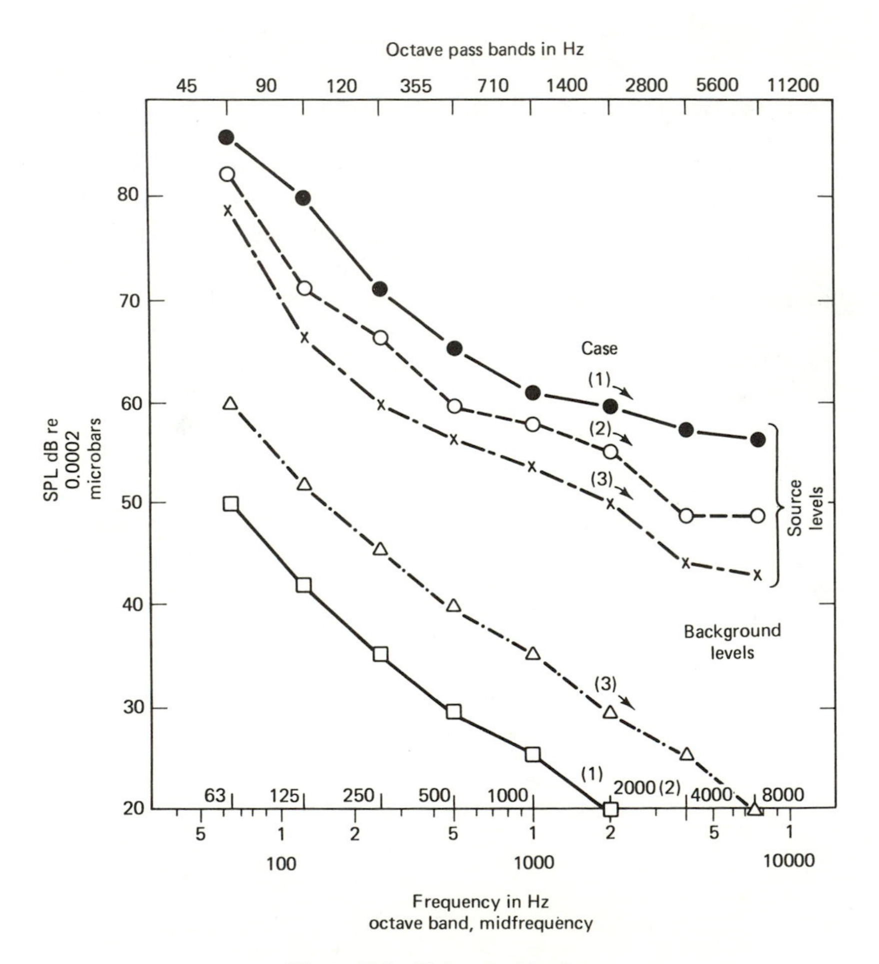

Figure 10.2 Three case histories.

that municipal authorities forced the facility to shut down after only 2 days of operation. The facility was allowed to run once more for acoustic testing and then was permanently closed until noise control devices were installed. In this case, the operation was in winter, the character of the noise was continuous, and there was no history of previous exposure.

Case 2 concerned a particular industrial noise source where the response was in the form of threats of injunction. These threats were removed after the installation of substantial noise suppression devices. In this case, the character of the noise was marked by predominant pure tones of sound. The operation was in the summer and there was a history of previous exposure within the residential area.

Case 3, with its background level 10 dB higher than the previous two cases, refers to a facility consisting of a loading platform with trucks, people shouting, horns, and so on. The response consisted of vigorous complaints to management, threats of legal action, followed by a court imposed injunction limiting the operation of the facility to a maximum number of vehicles and to daytime operation only. In this case, there was no history of previous exposure nor any identifiable pure tones.

To repeat, these three cases illustrate the importance of the intrusion factor of noise into existing areas. The greater the intrusion, the greater can be the expected community response to the intruding noise source.

10.7 GUIDELINES FOR THE APPLICATION OF FIGURE 10.1

The following summary can serve as a general guide to the application of Figure 10.1 for five types of neighborhoods, each with four assumed typical noise conditions. The numbers in the righthand column indicate the number of decibels of shift *upwards* which must be made to have the entire set of curves in Figure 10.1 applicable.

Description	Add dB
Very Quiet Suburban—Residential	
Continuous—night	0
Continuous—day	10
Peaking only—night	5
Peaking only—day	15
Suburban—Residential	
Continuous—night	5
Continuous—day	15
Peaking only—night	10
Peaking only—day	20

Description	Add dB
Urban—Residential	
Continuous—night	10
Continuous—day	20
Peaking only—night	15
Peaking only—day	25
Urban—Near Industry	
Continuous—night	15
Continuous—day	25
Peaking only—night	20
Peaking only—day	30
Heavy Industrial	
Continuous—night	20
Continuous—day	30
Peaking only—night	25
Peaking only—day	35

10.8 PROCEDURES FOR COMMUNITY NOISE SURVEYS

Prior to the establishment of any sections of a zoning act dealing with any type of community noise, it is necessary to take certain steps in a sequential and logical order.

Assuming that the normally used zoning descriptions are in existence, the first step is to review all areas of the community and particularly the boundaries of the zones to determine which shall need and receive the most preferential treatment with respect to varying degrees of required protection from noise. It is necessary to establish from various community contracts, which noise sources, at what times, and at what levels are responsible for complaints in various types of neighborhoods.

Having established the areas of concern, the next step is to determine on the basis of filed surveys, what the existing sound levels (that is, the ambient conditions) are in each of the particular areas of concern. This must be done for both daytime and nighttime periods, and should compensate if necessary for winter or summer operation. Here the minimum equipment required is a precision sound level meter attached to a high-fidelity, battery-powered tape recorder capable of long duration field recording. The recording, on playback, should be capable of presenting data which clearly delineates the variation in dBA levels over a typical 24-hr period. The data should indicate the maximum levels that occur and that time of day that they occur. The tape recording should be capable of laboratory analysis for other information such as octave band levels, one-third octave band levels, or statistical distribution of the noise background over different time periods. It is of utmost importance that the permanent record as provided by the recording, establish firm baseline reference levels for future comparison in the evaluation or judgment of the noise abatement goals.

As a third step in the establishment of performance zoning, a comparison of the measured levels obtained as above should be made with statistical norms drawn from national surveys and available from many sources to determine any anomalies for the particular community. For example, some communities have only a 5-dB difference in background levels from day to night while others have 10 dB or more. These differences may be townwide or merely confined to certain zones. Major highways or airports may seriously affect the community response to their respective noise levels.

Existing land use should be reviewed based on the survey data and analysis as the fourth step. Since the background noise levels can be utilized for the prediction of building occupant acceptance, it may be necessary to alter existing land use codes or draw new ones to comply with the probability curves for complaints such as those shown in Figure 10.1. The data obtained from the field survey will provide a practical base for the establishment of workable zoning regulations or ordinances for the control of potentially noisy activities or operations in or adjacent to various types of neighborhoods taking into account the character of the community. The data will provide a logical base for the siting and acoustic design of new structures and operations being introduced into an established community. Such structures might include new electrical substations, sewage, water pumping or refuse disposal stations, apartment houses, drive-in restaurants, or concert halls. The data can be used to anticipate the possibility of community complaints as a result of a change in operation in an existing facility. Such changes might include daytime operations of an industrial plant being extended into nighttime operations, the lengthening of an airport runway, or the widening of a main traffic artery to handle a larger volume and variety of traffic.

Finally, based upon the above steps, a list of fair and equitable spectrums can be prepared with representative numbers, unique to the community, which can then be written for a proposed code which will have the dual facet of being unique to that particular area and at the same time be compatible with national norms recognizable by any professional in the field.

10.9 NOISE BUFFER ZONING

The principle of noise buffer zoning is a highly effective but little used method of ensuring the least community response to objectionable noises. The principle can be simply stated. All high level noises should be grouped together with areas of lower level noises surrounding the core. The principle is illustrated for the general case in Figure 10.3.

Applications of the principle are shown in the following three figures. Figure 10.4 indicates an example of the treatment of a major highway intersection. Continuous interstate highways through rural areas are treated similarly. Figure 10.5 illustrates the successful method of laying out a research park along a major highway and at the same time providing for light industry. Figure 10.6 illustrates the special problem around airports which lends itself to the application of the buffer zone treatment. The FAA recommends a 3-mile distance off the center line of major runways as the minimum approach or encroachment of residential developments.

Violations of the buffer zone principle can cause a rise in the complaint level as regards noise. For example, placing a commercial zone, say a shopping mall, in the middle of a residential (A) zone can give rise to serious complaints and possible legal opposition unless stringent measures are taken to control the noise sources of the commercial establishment. This situation can be cured if anticipated at the design stage for a minimum of additional cost but becomes almost unsolvable after the fact of completed construction.

Arranging the zones in a progressive manner from most noisy to least noisy provides a mechanism for defining maximum noise levels not only within the zone but at the boundaries of the zones where the levels can be adjusted to avoid excessive hardship on the higher noise maker and at the same time provide adequate protection for the zone of lower noise level.

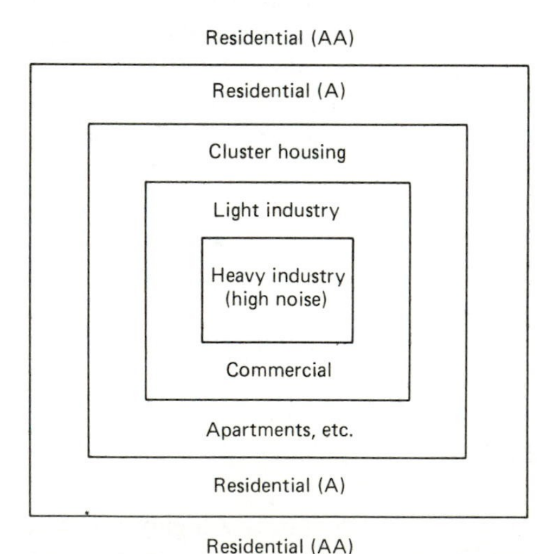

Figure 10.3 General principles of noise buffer zones.

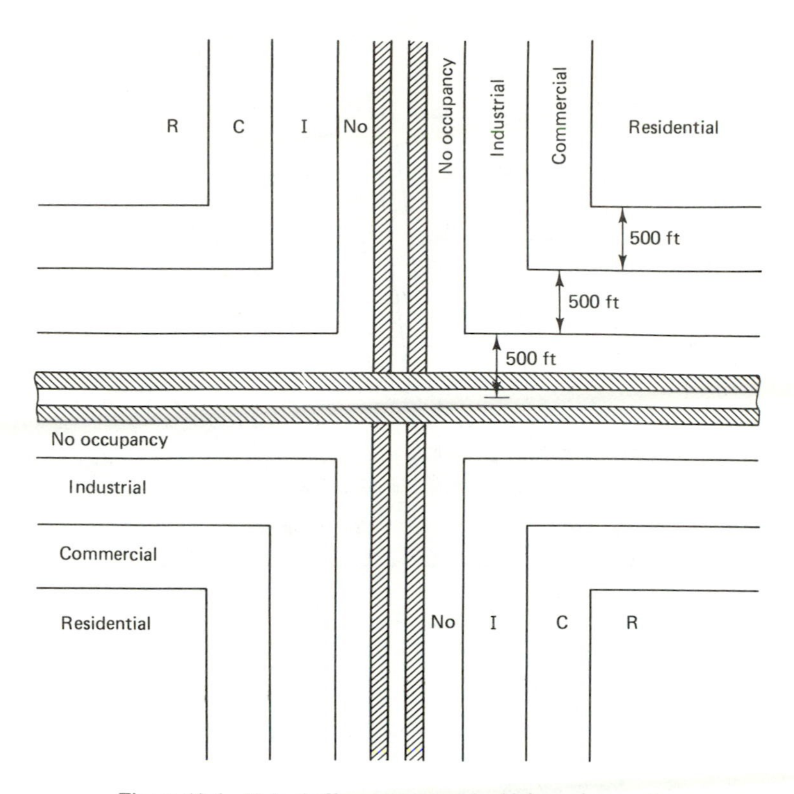

Figure 10.4 Noise buffer zones around a highway intersection.

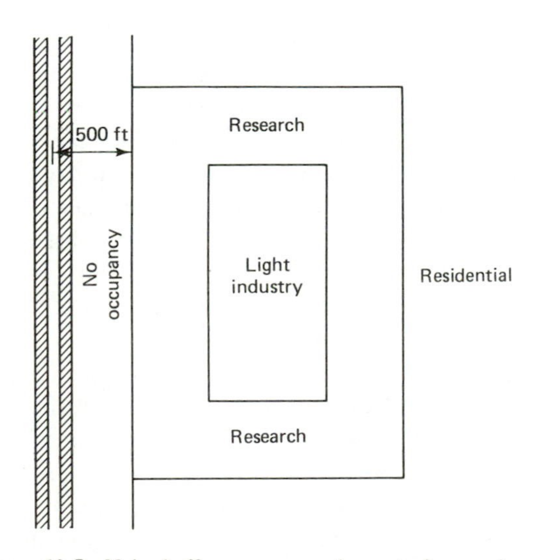

Figure 10.5 Noise buffer zones around an actual research park.

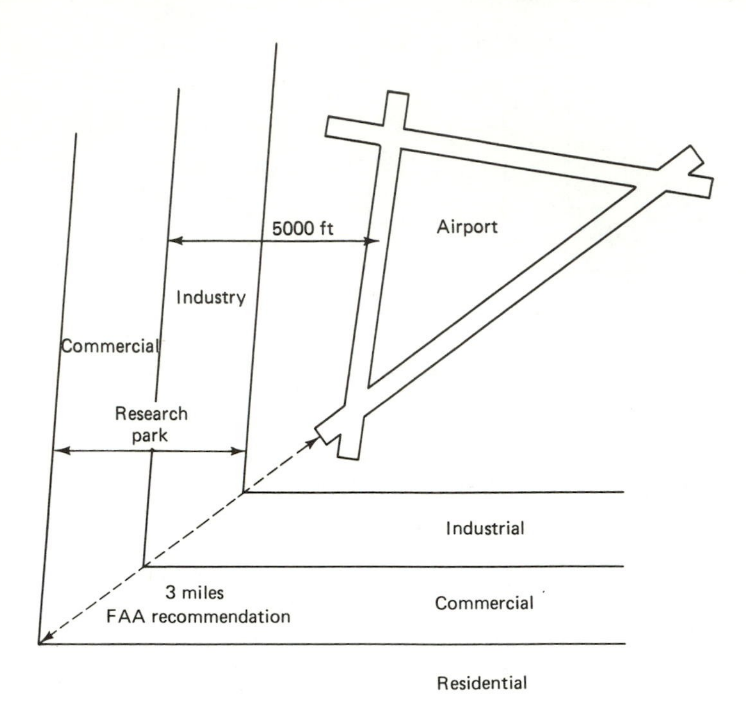

Figure 10.6 Noise buffer zones around an airport.

10.10 LEGAL ENFORCEMENT

A review of past and contemporary history of the successful control of noise by use of either the ordinance method or the zoning process reveals that the chief obstacle to success is the reluctance of law enforcement officials to take on this additional duty. Much of the reluctance is attributed to the fear of inaccurate measurements or the general lay concept that the subject is "too technical." Since enforcement is the heart of any successful control technique, the agency charged with this must be convinced of their ability to acquire adequate data. For this reason measurement techniques for checking a potential noise problem must be made simple and foolproof.

In general, whether or not the zoning regulations of noise ordinances list the required noise limits by octave band, one-third octave band, L_{dn}, or so on, those limits must be correlated as near as possible to readings on the dBA scale of a survey type of sound level meter. For most cases of community noise, this is a rather simple procedure, and was explained in previous chapters. In addition, *single event* criteria limits should be specified so that the enforcement officer can be alerted to potential long-term problems.

If the regulations call for specific levels at a specified distance from the source, say 50 ft, as read on the dBA scale, then the enforcement officer merely needs to set the equipment at that distance and read the sound level meter. If the regulations relate to motor vehicles, then again at a specified distance from the highway, say 50 ft, the peak level created by the vehicle during passby need only be observed and recorded. Construction site noise can be similarly measured.

If a violation for other than moving vehicles is noted by the enforcement officer, then and only then is it necessary to consider more elaborate measurement techniques. At this point, unless the enforcement officer has had special training in acoustic instrumentation and measurement techniques, it is expedient for all concerned that trained personnel, say an acoustic consulting engineer, be employed for the final determination of a violation.

The use of sound measuring equipment by law enforcement personnel is no more complicated than the use of radar for speed violations. The acceptance of readings taken by law enforcement officers for purposes of prosecution is slowly being accepted by the courts. The history of the acceptance of sound readings is sure to follow the history of the acceptance of radar speed readings where such acceptance on a universal basis has taken as long as over 10 years in some legal jurisdictions.

Communities need not make large investments in complex equipment. The purchase of precision-type sound level meters represents the main cost, a figure of approximately $800 in 1981. Less expensive calibrated instrumentation is readily available. As a rule of thumb, one piece of measuring equipment per 30,000 population appears to be adequate for those towns or states which already have effective law enforcement programs involving either ordinances or zoning regulations.

10.11 ENFORCEMENT BY ORDINANCE OR ZONING

The newest approach to the control of community noise is the enactment of new forms of legislation dealing with noise ordinances or model noise zoning. While the introduction of anti-noise laws is not a new concept, the present set of regulations makes use of principles not considered in the past.

The first principle is the use of the dBA scale as a measure of human response to objectionable noise. The reliability of this scale as a predictor of individual reactions is becoming internationally unquestionable. Thus it is possible to develop more satisfactory regulations dealing with adverse noise effects.

As proposed by the federal Environmental Protection Agency and adopted by many states, the new regulations make use of basic constitutional rights and obligations. The first right involves the principle that a person's home is his or her castle. Thus an individual has the right to carry on any activity within the confines of that individual's property. The right to make noise is recognized. But, on the other side, that same individual has an obligation not to disturb neighboring property owners. The obligation to refrain from becoming a nuisance is present. What is a nuisance and under what circumstances does noise from a neighbor become a nuisance is a basic question which is resolved to some extent by the new form of noise regulation. This type of regulation applies to stationary noise sources.

The new form of the laws or regulatory statements expressly uses what have been referred to as boundary line standards. They also make use of traditional land use definitions. Thus, there are boundary line standards between industrial, commercial, and residential properties. Maximum noise levels expressed in dBA are defined for each boundary. Exceeding these levels constitutes prima facie evidence that a nuisance in fact exists. For example, while a given industry may make as much noise as it desires on its own property, it is restricted to a maximum level in dBA as measured at its property line depending upon the use of the neighboring property. If the neighbor is another industry, the restriction is minimal. The restriction becomes more severe if the neighbor is rated as for commercial use, and the restriction is most severe if the abutting property is residential. The restrictions apply equally well in what might be called the reverse direction. A residential emitter is restricted as to the noise level which can be emitted to an abutting residence but is less restricted if the abutter is in a commercial or industrial use zone.

While the above process is being followed in many states and local communities, it is well to recognize that there is a significant problem associated with this process. The establishment of proper, enforceable noise levels in dBA in such a way as to distinguish between what noises are considered nuisances and what are part of everyday living becomes the major problem. States are beginning to adopt model legislation setting maximum limits for the various boundary conditions. These maximum limits are usually those compatible with the generally higher background levels of the cities. The model legislation usually permits suburban and rural communities to adopt more rigid standards to meet their own needs.

A word of caution must be introduced at this point. The establishment of fixed noise level limits should be preceded by a rather comprehensive, on-site noise survey. In this way the regulations can reflect the actual in-field conditions. The worst mistake that a community can make is to adopt numbers from another community's ordinance or zoning regulation. On-site measurements are mandatory for effective enforcement.

The choice of whether or not the noise regulations are encompassed in zoning regulations or are adopted as an ordinance is one that is made by local authorities. The question to be resolved is related to the choice of an enforcement mechanism. If the enforcement is to be done by police personnel, then the usual route is the enactment of a local noise ordinance. If the enforcement is to be through a zoning inspector, then, of course the regulations should be incorporated under the existing zoning code. Some municipalities have chosen a combined route to enlarge the scope of enforcement personnel.

Recognizing that certain noise sources either cannot be or should not be controlled, every set of regulations should include a section on exemptions. Typical exemptions are such sources as emergency vehicle sirens, church bells, unreinforced human voices (amplified sounds are not exempt), and construction noises if the construction is carried out between specified times of day and days of the week.

Also exempt are moving sources such as aircraft, which are solely regulated by the Federal Aviation Administration, and motor vehicles which are usually controlled by state regulations. The latter, however, are usually enforceable by local authorities with police powers. State regulations usually specify maximum noise emission levels as measured at some specified distance from the highway. Different levels are applied to three categories of sources (automobiles, trucks, and motorcyles) and to at least two speed limits for each of those vehicles. National standards have been proposed with respect to the noise limits of each vehicle, but only a few states have aodpted these standards.

PROBLEMS

For all problems, the following data is applicable: Complaints of the noise from a power plant have been received from the neighboring rural community. Most of the complaints have been about night operations. The octave band readings in the community register the following sound levels:

Mid-Band Frequency (Hz)	63	125	250	500	1K	2K	4K
Sound level (dB)	68	60	58	50	45	45	42

10.1 Considering the 90% probability curves in this chapter, what type of action might be predicted from the residents of that community?

10.2 If the neighboring community consisted primarily of light industry, what would be the predicted response of the residents?

10.3 Suppose in both of the above cases the measured noise only occurred in the daytime. How would this alter your answers?

10.4 Using material from previous chapters, determine for the measured spectrum its loudness level and its dBA level. Would you consider these levels objectionable? Explain.

11

Air Distribution System Noise

11.1 INTRODUCTION

The design of quiet air distribution systems for modern buildings has become an essential consideration for mechanical engineers. For this reason, it has become increasingly more important to evaluate the *noise components* of the air distribution system. The problem of evaluating and controlling the noise of these components involves all phases of the field of applied acoustics. Unfortunately, in many manufacturing fields, unless there are standard codes accepted by the industry and founded in engineering fundamentals, the noise ratings of a product are often strongly influenced by sales departments. Since an air distribution system is a collection of products designed to efficiently distribute the required air at predetermined temperatures and velocities with a minimum of energy loss through back pressure, sales literature can seriously affect the choice of components and often becomes the controlling element in the success or failure of the entire system. Instead of evaluating the performance of a product properly, ratings are tailored to that information which will best sell the product, tending to be colored by the practical necessity to sell under any conditions. This is particularly true when the noise rating systems involve single numbers which do not relate to the operating characteristics of the product. Sales departments play the numbers game delighting in disseminating information, for example, that their Product *X* has a 1-dB lower rating than their competitor's Product *Y*. They display their total ignorance of the fact that 3 dB represent the just noticeable difference in the judgment of a trained listener so 1-dB differences have no validity.

As another example, an automobile rated at 350 hp is commonly represented by the salesman as a *better* car than one rated at 315 hp, yet neither car will develop anywhere near this rated hp except perhaps on a dynamometer stand. Similarly, automobiles whose interior noise levels are purported to be measured at 85.5 dBA are claimed to be superior to those whose levels are measured at 86.2 dBA. Neither of these single number ratings truly describe the operating or comfort superiority of one car over the other. As has been indicated previously, energy in the low frequency region does not affect the dBA measurement scale of a sound level meter. Since the human ear is sensitive to a wide range of frequencies, the single number of a dBA reading, for all of its value in general terms, cannot adequately measure the *total* effect of sound on human comfort.

The air distribution industry (covering all phases of the air handling portion of heating, ventilating, and air conditioning systems) has found itself in much this same position. Air distribution devices rated at 36 dBA are purchased, or specified by the unsuspecting engineer, because it is presumed that such devices are quieter, and thus better products than those rated at 38 dBA. The fact that the 36-dBA device may have major low frequency components is neglected, or concealed, or otherwise not brought to the attention of the designing engineer. This will, in most cases, produce unsatisfactory if not disastrous results.

11.2 A TYPICAL AIR DISTRIBUTION SYSTEM

Figure 11.1 represents a schematic diagram of a typical air conditioning system combining the major components of both a high and low pressure system. The noise problem in low or high pressure systems differs only in magnitude. Effects on the noise level output produced by each component, with variations in static pressure and air volumes, will vary in magnitude for every air distribution system, but the basic problem is the same. The components selected in Figure 11.1 have been selected to demonstrate the sources of airborne systemic noises as one passes through a typical duct layout. Note that structure-borne noise is a separate problem for which current information is adequate to establish where vibration breaks should be used and to permit selection of proper vibration mounts for all of the air handling equipment. Remember that lack of attention to the potential paths for the transmission of vibrations can negate all efforts at reducing the airborne sounds. Design efforts in these cases can be both frustrating and fruitless.

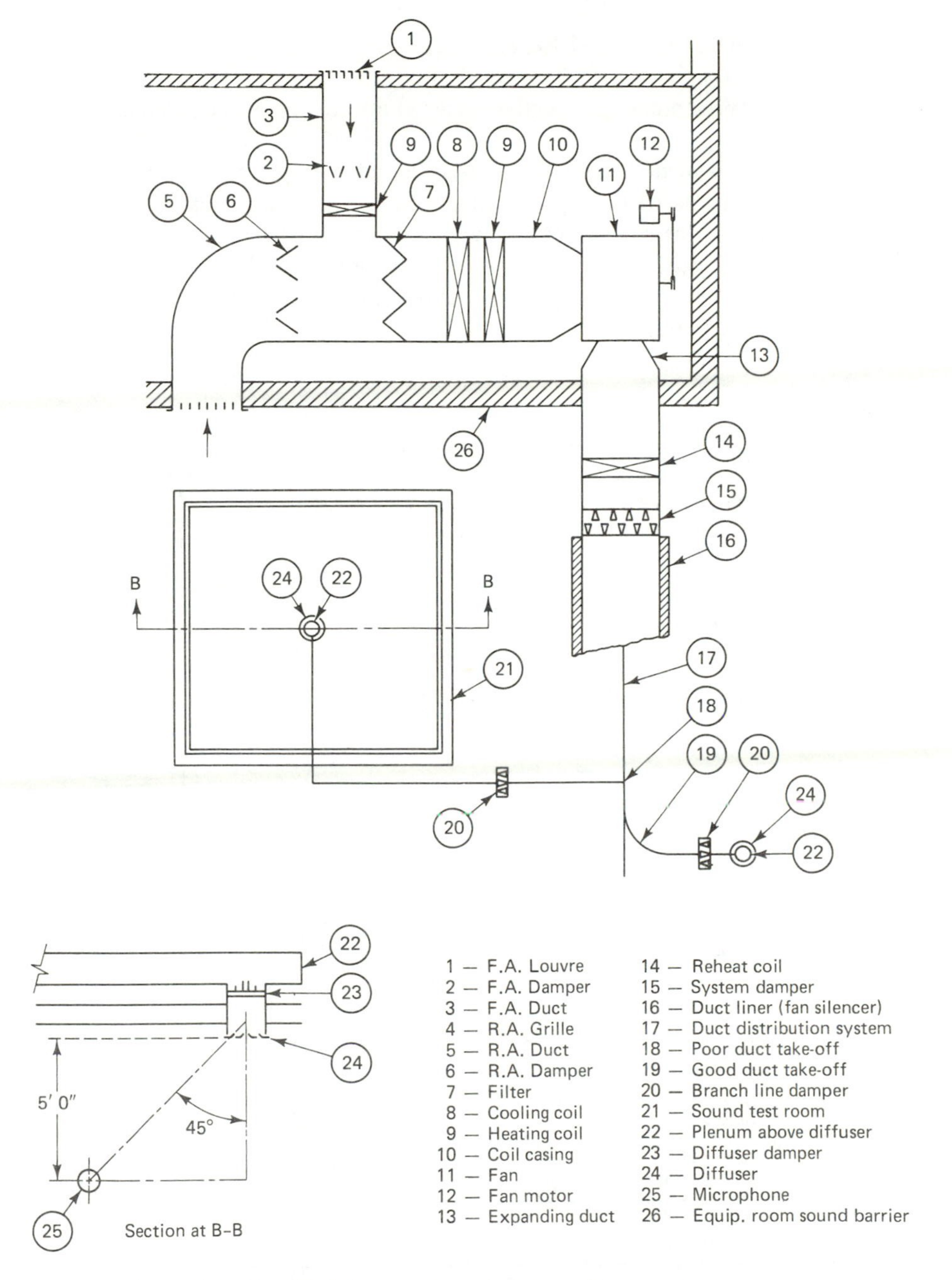

Figure 11.1 Simplified diagram of a typical air conditioning system.

11.3 RELATIVE ENERGY LEVELS

To obtain a proper perspective on the contributions of each element shown in Figure 11.1 to the output noise level, refer to Figure 11.2 where an attempt is made to show the relative sound energy contribution of each of the elements of the system. Increases in bar graph length indicate the addition of new noise sources while decreases indicate the effect of attenuating devices. The narrowing of the width of the bar graph indicates a tendency toward either the production of or the attenuation of single frequency components of noise.

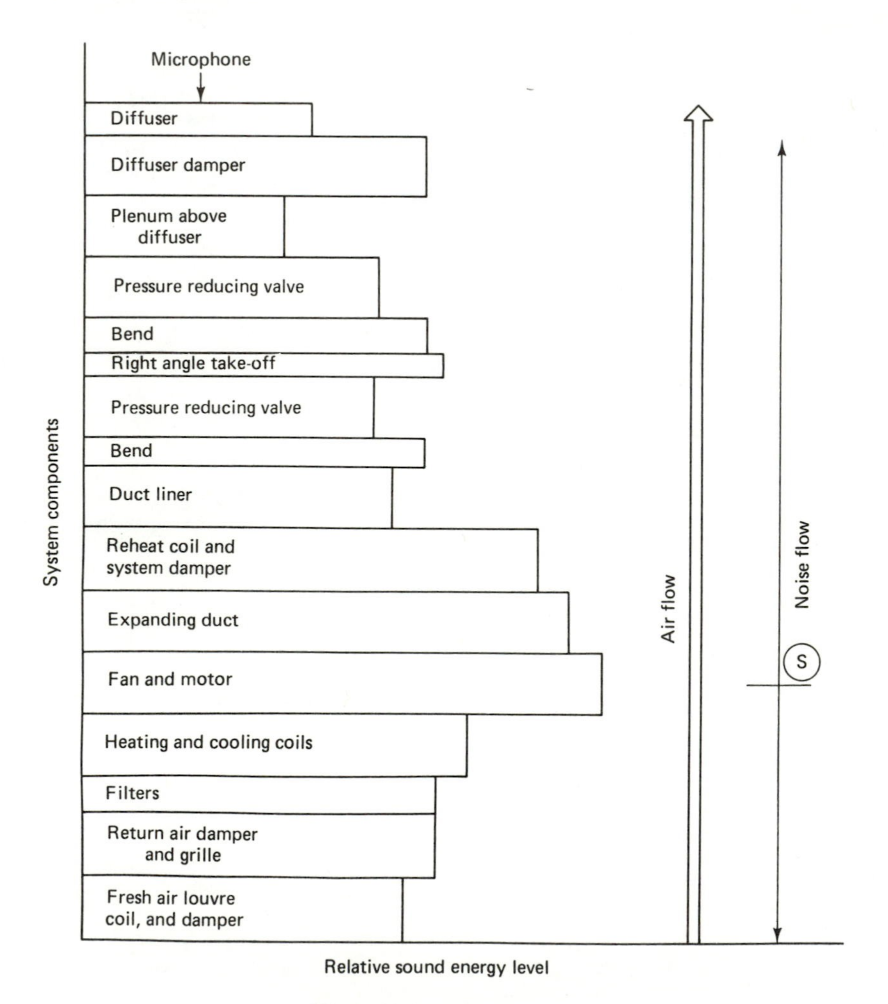

Figure 11.2 Systemic noise.

The chief source of the noise originates at the fan and motor and the noise is free to travel in both directions, toward the inlet (that is, upstream) as well as toward the outlet, in the system. A well silenced system is the result of a careful, although sometimes exhaustive, balancing of the additive (or noise producing) effects as well as the deductive (or noise absorbing) effects of the components in the system. Figure 11.2 can be used as a form of checklist for the designer to be used to acquire specific performance data on each proposed component. For example, the fan and motor, although the chief source of the noise, are often the most neglected areas in the system design. This applies both to the noise transmitted through the duct system as well as to the noise transmitted through the structure or through the walls of the fan room. Data for both paths is available from the fan manufacturer and should be obtained.

11.4 FAN ROOM NOISE

The major sources of noise in an air distribution system emanate from the fan room. In this area, outside noises can enter the system through fresh air intake louvers, noises from within occupied areas of the building can reenter the system through return ducts, and equipment contained in the fan room such as refrigeration compressors, condensate pumps, fans, and fan motors add noise to the air entering the distribution system. In the design of each of these components, manufacturers must give careful consideration of their sound producing characteristics and produce equipment with the lowest possible emission characteristics.

It is not sufficient that fan rooms be located as far as possible from potentially noise sensitive areas in the building but the room itself and the equipment it contains need special consideration. In addition to avoiding attaching noisy or vibrating equipment directly to structural components of the building by providing adequate vibration isolators, all pipe and conduit connections to the machines must be similarly isolated. Rigid pipe or conduit connections serve as excellent bypass paths for the transmission of sound and vibrations to the adjacent structure.

The walls, floors, and ceilings of the fan room must have the characteristics provided by high transmission loss structural components (see Chapter 5). In addition, to provide for the occupational safety (hearing protection) of the personnel required to work in this area, walls and ceilings should be treated with acoustically absorbent materials to reduce the reverberation characteristics and provide for some sound reduction through the absorption mechanism. See Chapter 4. An overview of these problems is indicated in Figure 11.3.

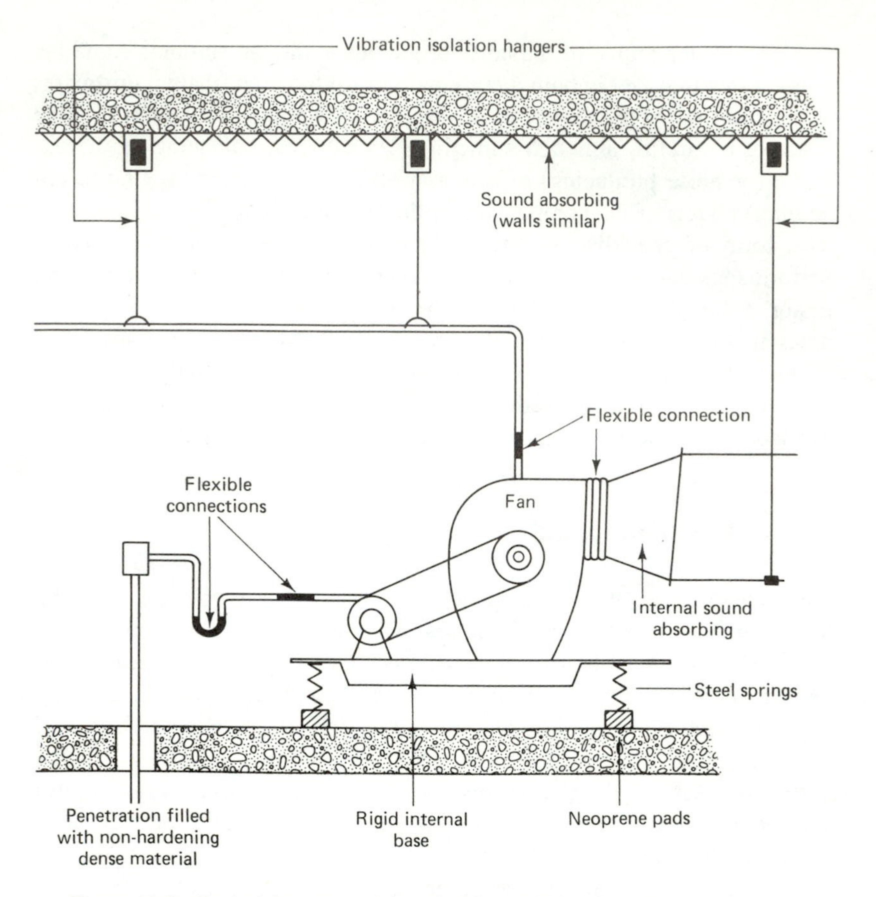

Figure 11.3 Typical fan room. (Adapted with permission from *Concepts in Architectural Acoustics,* by M. David Egan, McGraw-Hill, 1972.

11.5 FAN NOISE

The fans which are found in buildings are usually classified into two types. Axial flow fans with propeller-type vanes have their air flow parallel to the axis of rotation and generally operate against very low static pressures. Centrifugal fans can be subclassified into tangential or blade types. In either case the air flow makes at least one right angle bend through the unit. Typical sound spectrums of each of the above types is shown in Figure 11.4. Both types of fan emit a broad-band-type of noise with some discrete tones being more prominent than other frequencies. The axial-type fan has fundamental and harmonic tones directly related to the number of blades and the speed of rotation of the fan as follows

$$f_n = \frac{nB_nR_s}{60} \tag{11–1}$$

where f_n is the frequency of the nth harmonic

B_n is the number of blades

R_s is the rotational speed in revolutions per minute

Another discrete tone plus its harmonics can be generated air flow past either supporting struts or stationary diffuser struts. If the number of blades is designated as B_n and the number of struts as B_s, then a discrete tone will occur when the following frequency relationship holds

$$f_S = \frac{nB_nB_SR_S}{60\ K} \tag{11–2}$$

where K = the highest common number present in B_n and B_S. Thus, if the fan has three blades and its 1800-rpm motor is supported by six struts, the value of the fundamental frequency due to the struts would be

$$f_S = \frac{1(3)(6)(1800)}{(3)60} = 180\text{ Hz}$$

This frequency would occur in addition to the fundamental blade passage frequency [from Equation (11–1)] of 90 Hz and would match and strengthen the first harmonic of the blade passage frequency at 180 Hz.

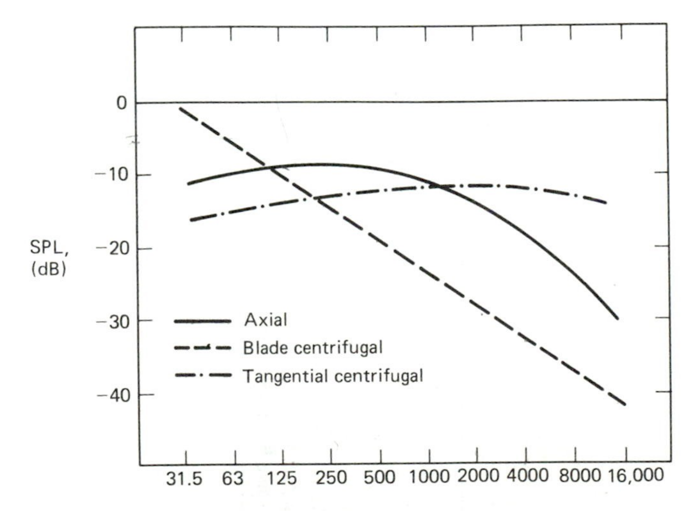

Figure 11.4 Typical spectrum diagrams for axial and centrifugal fans.

	Octave Bands									
	31.5	63	125	250	500	1000	2000	4000	8000	16,000
Total PWL	111	111	111	111	111	111	111	111	111	111
Correction from Figure 11.4 (−)	11	10	9	8	10	12	15	18	22	30
Predicted Octave Band Spectrum	100	101	102	103	101	99	96	93	89	81

In addition to the discrete tones due to the blade passage, the axial flow fan creates broad-band aerodynamic noise due to vortex shedding from the blades. Reducing the fan speed (rpm) or reducing the vortex shedding by blade redesign will have a significant effect on the output noise of the fan.

As an engineering rule-of-thumb it is possible to approximate (within 3 to 4 dB) the output sound power level (RWL) of an axial fan by use of the following equation

$$\text{PWL} = 10 \log \text{hp} + 10 \log p + 90 \qquad (11\text{–}3)$$

where hp is the motor shaft horsepower and p is the static pressure drop in inches of water. Using this value, a spectrum can be predicted for octave bands by use of Figure 11.4, achieving the correction factors for each octave band from the axial fan curve of that figure. The following is an example of the process.

Example:

Given a 15-hp axial fan producing an 8-in. H_2O static pressure, estimate the octave band power levels.

$$\text{PWL} = 10 \log 15 + 10 \log 8 + 90$$
$$= 111 \text{ dB}$$

Using the corrections from Figure 11.4 we set up the table which is shown on page 160 of this text.

The noise from centrifugal fans is a combination of discrete tones at the blade passage frequency and aerodynamic noise from the shearing action of the blade and the resultant turbulence. The sound power level can be predicted by using Equation (11–3). The spectrum can be predicted in the same manner as is shown in the above example by using the proper centrifugal curve of Figure 11.4.

11.6 OTHER FAN ROOM NOISE SOURCES

Another major noise source within the fan or machinery room is the compressor. Again these units fall into two major classifications: axial flow and centrifugal compressors. While noise levels from each type can vary considerably from manufacturer to manufacturer, thus emphasizing the necessity for obtaining specific data from the builder of the equipment, it is possible to estimate the sound power level for each type. For axial flow compressors the relationship is as follows

$$\text{PWL} = 20 \log \text{hp} + 80 \qquad (11\text{–}4)$$

where hp is the rated horsepower of the compressor. In order to estimate the sound power level for centrifugal compressors, Equation (11–4) is modified as follows

$$\text{PWL} = 20 \log \text{hp} + 80 + 50 \log \frac{B_v}{800} \qquad (11\text{–}5)$$

where B_v is the blade tip velocity in ft/s.

Motors and feed pumps can also be major contributors to the noise level of a fan or machinery room. Data on these units is so varied that design information should be obtained from the manufacturer. Most manufacturers rate their units in dBA at distances of 3 to 5 ft. The levels can range from 63 to 91 dBA depending on the manufacturer as well as the type and horsepower of the motor or pump unit. Thus a serious error in the analysis of expected noise levels within the fan room can be made if these sources are not accounted for.

11.7 DUCT SYSTEM NOISE SOURCES

Dampers for controlling air flow, distribution terminal boxes for distributing air from main lines to branch ducts, and grills, registers, and ceiling diffusers are all potential noise sources within the duct system. Under controlled conditions, these same devices may also provide for attenuation of the airborne noise passing through the duct system. While this seems like an anomaly, it can be explained by examining the critical design factors in each component.

A large portion of the noise in a duct system can be controlled at the design stage by careful attention to some of the standard design components. Figure 11.5 illustrates some of the more common noise sources in the duct system. In each case, the aim is to reduce the potential for the creation of turbulence in the air stream to a minimum. Turbulence is a source of noise. Right angle bends, branch take-offs, and diffusers should be designed with rounded corners, while control valves which have aerodynamically smooth contours are quieter than those with sharp edges. Covering a duct with insulating material provides no attenuation of the sound along the duct passage but simply supplying a liner material will allow the sound to be radiated through the sides of the duct. A combination of liner and wrapper provides the optimum treatment. Grills, registers, and diffusers should be sized for a minimum of air flow since these units are inherently noisy. The noise levels are a function of the face velocity.

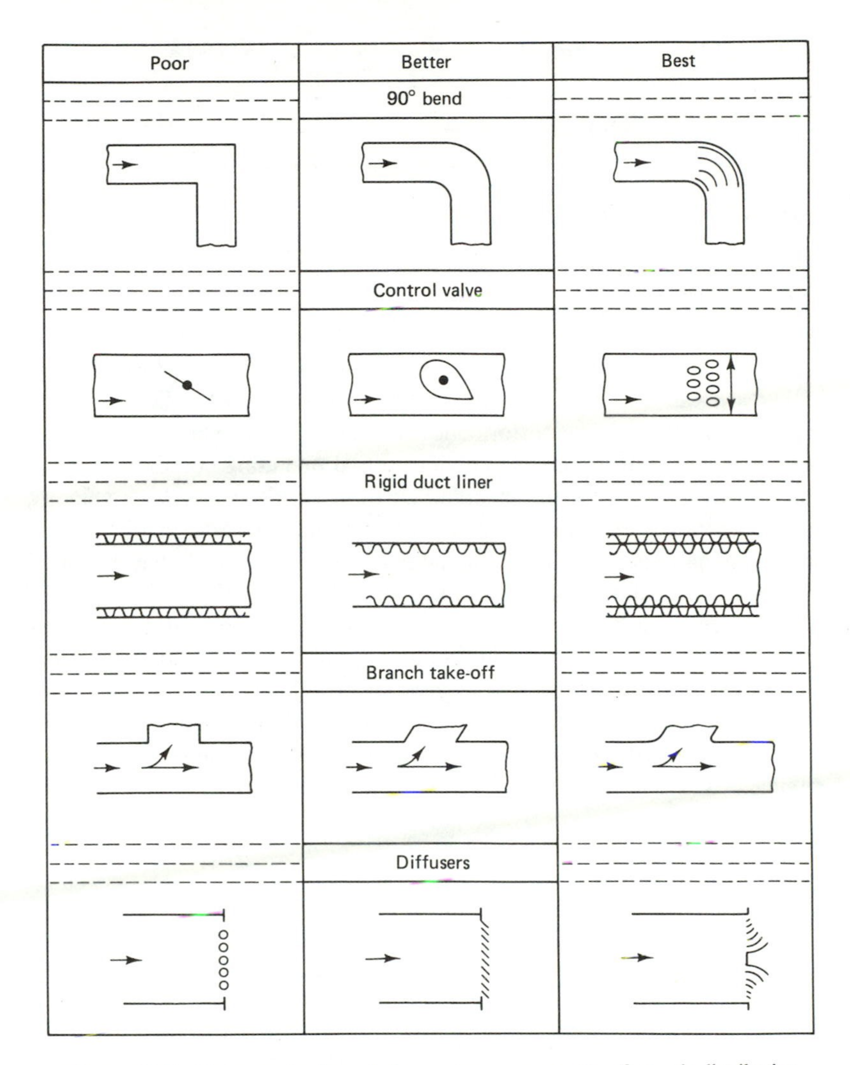

Figure 11.5 Noise evaluation of the system components of an air-distribution system.

According to some research results the sound output is a function of the face velocity to the fifth power and according to other researchers it is a function of the velocity to the eighth power. In any event, designing for the lowest practical velocity to provide for adequate air circulation in the receiving room can be of great benefit in producing a silenced system. As a guide for design purposes, Table 11.1 lists some of the recommended maximum jet or face velocities for grills or diffusers.

Pressure-reducing valves with an aerodynamic design can provide

TABLE 11.1 Recommended Maximum Jet or Face Velocity for Diffusers or Grilles

Grille or Diffuser Application	Face Velocity (ft/s)
Industrial	30
General offices and stores	25
Private offices and theaters	15
Residences, churches, and hotels	12
Broadcast studios	5–10

some attenuation characteristics as shown in Figure 11.6. Most of these valves are patented devices bearing the classification of PRV valves by their manufacturers. The performance of each type will vary slightly from the curve shown in Figure 11.6 under normal operating conditions. If the valve is sized wrong for the specified air flow (resulting in a high velocity across the valve) the attenuation will be lost and the valve can become a significant new source of aerodynamic noise because the turbulence created by the high internal velocity becomes very high.

Distribution boxes designed to distribute air to several adjacent locations from a main air duct can produce significant noise reduction if the box is lined with acoustically absorbent material. Similarly, mixing boxes designed to mix cold and hot air before it is distributed through a diffuser to the receiving space can serve as a part of the silencing system if the mixing box is acoustically treated with an adequately designed acoustic liner.

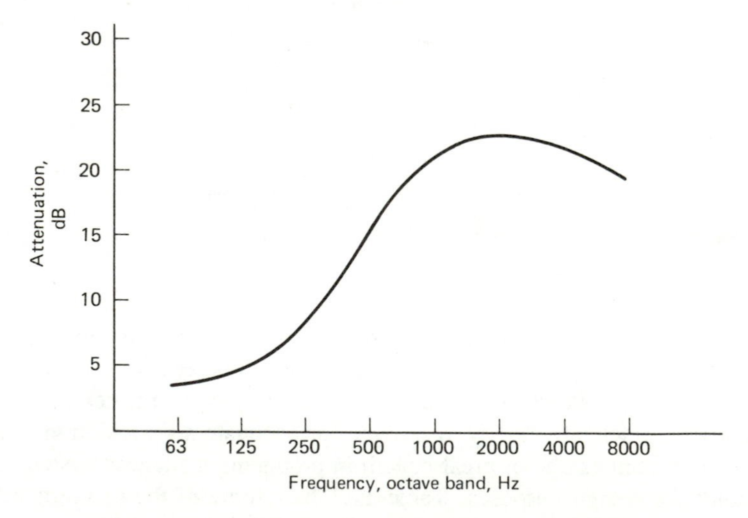

Figure 11.6 Attenuation of typical pressure-reducing valves.

11.8 ACOUSTIC TREATMENT FOR DUCTS

Lining the ductwork is generally accepted as the most economical means of quieting an air distribution system. The lining should begin at the plenums located just fore and aft of the fan. The thickness of the liner typically ranges from 1 to 2 in. and should be located on all four sides of the air duct. A useful empirical equation developed by Sabine gives a conservative estimate of the noise reduction achieved by lining a duct with a Fiberglas-type material (in English units)

$$A = 12.6\,\alpha^{1.4}\frac{P}{S} \qquad (11\text{–}6)$$

where
A is the attenuation in dB/ft
α is the absorption coefficient of duct liner for random incidence sound
P is the perimeter of the duct in in.
S is the cross-sectional area of the duct in in.^{2}

For round ducts, the above equation becomes

$$A = 50.4\,\alpha^{1.4}\left(\frac{1}{D}\right) \qquad (11\text{–}7)$$

where D is the duct diameter in in.

For plenums, an estimate of the reduction in sound in decibels can be obtained by using

$$A_f = 10 \log \frac{\alpha_A}{\alpha_B} \qquad (11\text{–}8)$$

where
A_f is the total attenuation of some frequency
α_A is the product of the value of α of the absorbing material at a specific frequency multiplied by the area covered by the material
α_B is the product of the α of the metal at a specific frequency times the area covered by the absorbing material

Thus if 10 ft^2 of absorbing material with an α at 1000 Hz of 0.7 is placed against a metal wall with an α of 0.05, the total attenuation to be expected at 1000 Hz is

$$A_T = 10 \log \frac{(0.7)(10)}{(0.05)(10)}$$

$$= 10 \log 14 \simeq 11 \text{ dB}$$

The above theory and application was developed for rigid duct, and the advent of flexible duct and its applications has led to further studies regarding the use of specially constructed duct for sound attenuation as well as for the more efficient and less costly conduction of air around narrow corners, sharp turns, and in small spaces.

Note that the sound attenuation of ducts can be tested either with air moving through them or under static conditions with no air flow. It seems obvious that a unit designed to conduct air should be tested under simulated field conditions; that is, with air flow. Surprisingly, the truth is that this was not the case until very recently. Many erroneous data exist in the literature concerning the performance of sound absorbing mechanisms (including not only data on duct performance but data on the performance of the packaged silencers which will be discussed later) because the tests were performed under still air conditions. Since these data do not include the effects of self-generated noise (that is, turbulence, which reduces the net performance of the attenuating device), they mislead through omission. And, in a very questionable way, they present a more buyable product to the unwary design engineer who is comparing duct performances.

Until 1973, when the Air Diffusion Council adopted and published Regulation FD-72, Subsection 3, no national code covered the evaluation of sound attenuation and generation in ducts with air moving through them. This code now provides a uniform method of comparing competitive flexible ducts as well as a model for testing rigid-duct configurations. The basic technique is the use of the *insertion loss* method for determining the acoustic properties. By this method, the difference between two sound pressure levels measured at the same point in space before and after the muffling device is inserted in the line becomes the measure of performance. Test results indicating the sound attenuating capabilities of the device are reported both with and without the effects of self-generated noise. As has been indicated before, the air velocity in an air conditioning duct significantly affects the amount of noise generated. The higher the velocity, the higher is the noise level. For example, turning vanes in sheet metal elbows will add about 7 dB more noise at 2000 ft/min than at 1000 ft/min. Diffusers perform similarly. Thus, it is very much the concern of the design engineer that any air conducting device such as a flexible duct be rated under specific air flow conditions. For this reason, FD-72 specifies test conditions involving at least four different air flows as well as the no-flow condition. The acoustic measurements are reported for at least seven octave bands with center frequencies ranging from 125 to 8000 Hz.

Sound transmission through the sidewalls of an air duct is often as important a factor in design as are attenuation or self-generated noise. When air ducts are located above a suspended or integrated ceiling, noise passing through the duct walls can be transmitted directly to the occupied space below unless adequate precautions are taken. Thus, FD-72 provides for measurement of duct wall transmission as an added requirement for duct rating. A duct may have good attenuating characteristics but a poor

TABLE 11.2 Performance Characteristics of Various Types of Duct

	Acoustic Performance	
Type of Duct	Attenuation or Net Insertion Loss	Transmission Loss—Side Wall
Rigid mental duct	Poor	Good
Coated fabric or film, Metal reinforcement, Uninsulated	Fair	Poor
Corrugated metal, Uninsulated	Poor	Fair
Coated fabric or film Metal reinforced core, Exterior insulation, and vapor barrier	Fair	Fair
Corrugated metal core exterior insulation and vapor barrier	Poor	Good
Corrugated perforated metal core, exterior insulation, and vapor barrier	Good	Excellent
Permeable air core with metal reinforcement, exterior insulation, and air barrier*	Excellent	Poor
No Core, Wire Supported Insulation and an Air Barrier*	Excellent	Poor

*These types are for low pressure and low velocity.

duct wall transmission loss rating. The system designer must determine, from test data available which duct qualities he or she must have, which can be lived with, and which cannot be tolerated. By carefully considering duct information and consulting a table such as Table 11.2, the engineer can make an intelligent and efficient choice.

11.9 PACKAGED SILENCERS

As an outgrowth of research into the mechanism for achieving attenuation of sound through acoustical lining of ductwork, as well as the need in many instances for achieving high attenuation values in relatively short distances, the development of the *packaged silencer* or *muffler* has occurred. The term "silencer" has been used in and associated with the air distribution industry as well as in the problems of silencing jet engine test cells. Muffler, on the other hand, has been commonly used to refer to the device required for quieting the exhaust of internal combustion engines (that is, automobile, diesel engines, power mowers, and so on). Thus, the use of a particular name is not because the principles of operation are different but merely due to the common reference in a particular trade.

The packaged silencer or muffler falls into one of two subclassifications depending upon the mechanism used to provide the attenuation of the sound. Thus, these silencers are classified as either absorptive or reactive. The performance of each type of muffler is usually given in octave bands for a range of air flows. Historically, many manufacturers have reported the results of their tests under static (no-flow) conditions and thus much misinformation concerning the in-place performance of the units exists in the literature. Since these units are highly sensitive to the phenomenon of self-generated noise (turbulence), only data taken under actual air flow conditions (usually obtained by the insertion-loss method) can be relied upon and used by the system design engineer. Both absorptive and reactive types of silencers occur in many physical configurations but the basic concepts of each type are unique.

Absorptive silencers, often referred to *parallel-baffle* silencers, attempt to achieve a high acoustic performance by splitting the air stream with aerodynamically designed baffles providing an exposure to a large amount of absorbing material in a relatively short distance. Each baffle (or divider) is usually constructed with perforated walls backed by highly absorbent acoustical material which usually has a base of glass or mineral wool. The acoustical performance of this type of packaged silencer depends primarily on the width of the air passage between the baffles (that is, the

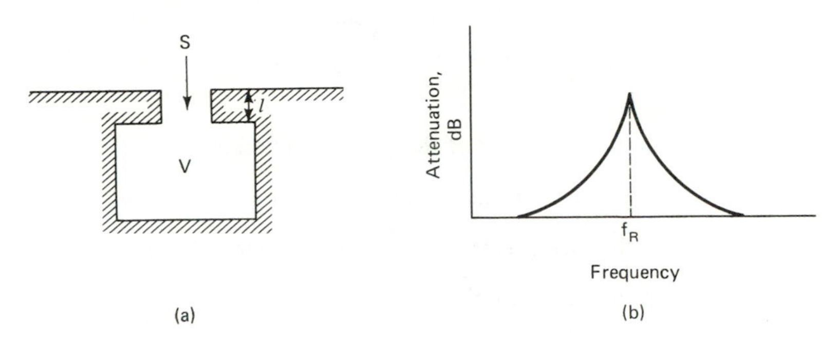

Figure 11.7 Helmholtz resonator.

spacing), the thickness of the baffle itself, and the length of the baffle. Whether or not the entire silencer is rectangular or circular in cross section, or whether or not the baffles are straight, staggered, or curved, the principle is the same, that of providing the largest possible area of sound absorbing material exposed to the passing sound wave. Note that while the acoustical performance increases with a decrease in the spacing between the baffles or with the thickness of the individual baffle, there is a considerable sacrifice in these design changes with regard to the back pressure developed for each of a manufacturer's configurations as well as for each of the rated air flows. A balance between this pressure loss and the attenuation which is desired is essential. This type of unit is essentially a high frequency type of absorber, being most efficient at those frequencies where the liner material is most absorbent, usually from 500 to 2000 Hz. Performance data for this type of silencer must be obtained from the manufacturer's tests since no reliable theoretical or empirical equations exist to predict the results with any accuracy at this time.

Reactive mufflers also take many forms and shapes but again the basic acoustic principles are the same. As the name implies, the performance of these silencers depends upon the energy losses occurring during reflection and expansion of sound waves as they impinge upon or pass by physical discontinuities in the flow path. Two simple types of reactive or dissipative mufflers, upon which the more complex types are developed, can be discussed.

The Helmholtz resonator shown in Figure 11.7 (a) is a classic case of a highly effective but limited to a single frequency (or narrow range of frequencies) type of reactive silencer as indicated in Figure 11.7(b). From the research of Helmholtz, whose name is attached to this type of silencer, the frequency of resonance can be approximated by the following equation which makes use of the physical dimensions of the muffler.

$$f_R = \frac{C}{2\pi}\sqrt{\frac{S}{l_n V}} \qquad (11\text{–}9)$$

where f_R is the frequency of the peak attenuation

C is the velocity of sound (corrected for temperature)

S is the cross-sectional area of the opening

l_n is the neck length

V is the volume of the cavity

This type of reactive muffler can be used successfully to dampen those noise sources which have a characteristic single frequency sharp peak in their noise spectrum. Several of these (with proper design adjustments) can be used in series if the peaks occur at other well-defined frequencies.

The other elementary form of a reactive muffler or silencer is the simple expansion chamber as is shown in Figure 11.8(a) with its spectral performance indicated in Figure 11.8(b). The acoustic performance of such a simple expansion chamber can be predicted by the following equation for transmission loss

$$\text{TL} = 10 \log\left[1 + \frac{1}{4}\left(m - \frac{1}{m}\right)^2 \sin^2 kl\right] \qquad (11\text{–}10)$$

$$\text{where } m = \frac{\text{Muffler cross-sectional area}}{\text{Duct cross-sectional area}} = \frac{S_2}{S_1}$$

$$k = \text{wave number} = \frac{2\pi}{\lambda} = \frac{2\pi f}{c}$$

and l = muffler length

Examining the above equation, we note that the attenuation vanishes when $kl = n\pi$, for values of $n = 0,1,2,3, \ldots$, and that the maximum attenuation occurs at values of $kl = \frac{n\pi}{2}$ for values of $n = 1,3,5 \ldots$. This explains the multiple attenuation peaks shown in Figure 11.8(b).

Manufacturers of the reactive type of silencer create many configurations for many different purposes by combining the concepts of the Helmholtz resonator and the expansion chamber. The combinations in many cases create broad-band attenuators which are often more useful at low frequencies than is the strictly absorptive type of silencer. Both absorptive and reactive muffler principles have been combined in other note-

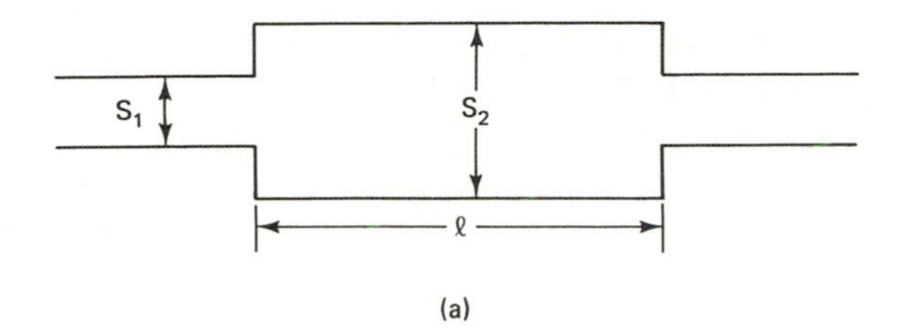

(a)

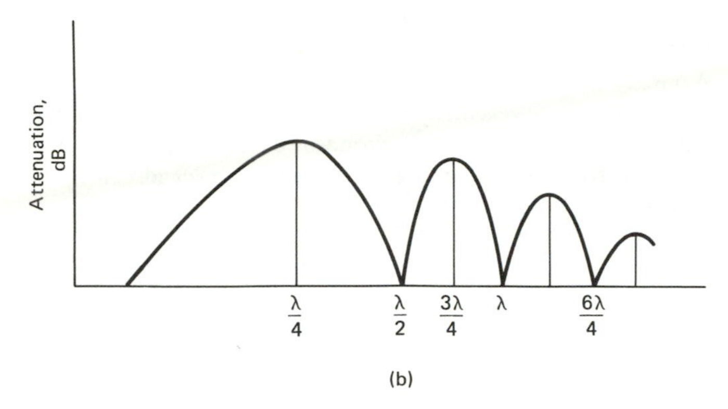

(b)

Figure 11.8 Single expansive chamber.

worthy cases such as test cell silencing mechanisms for jet engine noise. Reference should be made to manufacturers' literature for specific applications of either type of muffler.

PROBLEMS

11.1. A four-bladed axial flow fan is being driven by an 1800-rpm motor and is supported in its cage by three struts.

a. Determine the fundamental tone put out by the fan and its first four harmonics.

b. Determine the fundamental and first four harmonic frequencies due to the struts.

11.2. The motor shaft horsepower of the fan in Problem 11.1 is 5 hp and the static pressure drop equals 2 in. of water. Calculate the total sound power output and the frequency spectrum.

11.3. If the fan in Problem 11.1 is changed to a centrifugal fan with the same general characteristics (Problem 11.2), calculate its frequency spectrum.

11.4. The acoustical lining in an air duct has a random absorption coefficient of 500 Hz of 0.4 and at 1000 Hz of 0.7. If the duct dimensions are 10 in. by 12 in., calculate the anticipated attenuation for a 10-ft length of this duct.

11.5. Calculate the size of a round duct that will give the same performance as the rectangular duct of Problem 11.4.

11.6. A fan plenum is roughly 5 ft by 6 ft by 10 ft long. Acoustical material is to be placed on the sides of this plenum. The material has an absorption coefficient of 0.75 at 1000 Hz, while the metal wall has a coefficient of 0.1. Calculate the sound reduction that can be achieved by the addition of this treatment.

11.7. It is desired that a Helmholtz resonator be designed to have a resonant frequency of 300 Hz. Calculate the volume required if the opening is to be 4 in.^{2} and the orifice length is 2 in. Assume the temperature is at standard conditions.

11.8. A 6-in. diameter round duct enters a 12-in. diameter pipe which is 3-ft long and to which is attached another 6-in. diameter round duct. Calculate the transmission loss through this expanded section at 1000 Hz. What would happen if the expanded section were doubled in length? Show your reasoning.

12

Architectural Acoustics

12.1 INTRODUCTION

The field of architectural acoustics generally implies the mechanism of control of sound in auditoriums, music halls, or opera houses. The principles of architectural acoustics, in fact, applies to a much broader set of locations. They are equally applicable to factory spaces, commercial buildings, offices, or homes. Thus, wherever a closed space exists, where some form of communication (speech or music) is desired, the control of the sound environment is possible with the proper application of the principles of architectural acoustics. Control is exerted by a combination of the phenomena of reflection, diffraction, absorption, and reverberation time. All of these phenomena have been previously discussed in the opening chapters of this textbook. The art or science of combining these phenomena to exercise control of sound in closed spaces is a rather crude definition of the words "architectural acoustics." It is the aim of this chapter to present an orderly approach to the analysis of either an existing space or a new space using tools previously discussed.

12.2 SHAPE FACTOR

One of the first steps to ensure even distribution of sound in a given space is to use ray diagrams, as described in Chapter 2, to determine the multiple paths of the sound originating within the space. Scale drawings of the plan and elevation of the structure are most useful. After locating the source or sources, ray diagrams can be drawn in sufficient number to indicate potential problems such as standing waves (reflections from parallel walls) or focusing effects from curved walls or path differences sufficient to give an echo effect. Reflections between parallel walls will cause a flutter effect where the received sound appears to be rising and falling in intensity when in fact the source is at a constant level. Focusing from concave or parabolic surfaces causes an intense buildup of the sound in a specific location. Sounds from reflecting surfaces which arrive with a significant time delay over the path of the direct sound will cause the listener to hear echos, be unable to distinguish the meanings of speech, or hear music in a fuzzy, unclear manner. The purpose of the ray diagrams is to locate surfaces which create an adverse effect on the prime purpose of the study, to achieve a uniform sound level throughout the listening area.

12.3 REFLECTION

Methods of eliminating the destructive effects arising out of the ray diagram study are varied. For example, the elimination of the flutter echo caused by parallel walls is accomplished in new construction by making one of the walls tilted (in the vertical plane) or deliberately misaligned (in the horizontal plane). Building one of the walls in a rippled or shallow triangular section can be very effective and at the same time create an interesting architectural effect. Tilting either the back stage wall or the rear wall of an auditorium will achieve the same effect. In the case of existing spaces where wall configurations cannot be altered, it is still possible to eliminate the flutter or echo effects by applying sound absorbing material to those walls creating the problem. For example, the application of such materials as velour drapes, deep pile rugs, banners, as well as the standard acoustic tiles can be very effective in eliminating a hard sound reflective surface. Since the equations for reverberation control do not require the specific location of any absorptive material, the application of the above materials on the walls constitutes a portion of the required total absorption for the space.

12.4 DIFFRACTION

Diffraction effects are usually of no concern to the architect in the design of auditoriums. However, it is well to note that if there is to be a deep stage with a relatively small proscenium arch, the sounds emanating from the rear of the stage will not be heard very clearly in those front seats to the right and left of the arch. Those seats will be in the shadow zone caused by diffraction at the vertical edge of the arch. Reflective surfaces must be used to reflect sounds into these areas to assure uniformity of hearing conditions within the auditorium. In a similar manner, projections from the side walls will not usually set up any control due to diffraction. In order for the diffraction effect to be present, the object must be larger than the wavelengths of the sounds to be controlled. Thus, an obstacle presenting a face to the sound source which is only 1 ft wide will control frequencies above 1000 Hz (wavelength of 1 ft). Lower frequencies with longer wavelengths will not "see" the obstacle.

The adverse effects of diffraction must be recognized in the design of open-space offices or schools. The use of partial height barriers to separate activities may provide sight barriers but will provide a minimum of sound reduction. The fact that the sound will bend (diffract) over the barrier, creating only a limited shadow zone, is often neglected. If the space divider does not interrupt the line of sight between the source and the receiver, it will not provide any sound reduction.

The use of partial height barriers in industry to surround noisy machines is an equally ineffective mechanism to protect the workers from hearing losses. Partial protection for the operators can be achieved provided their location is within the shadow zone of the partition. Care must be taken that the partition does not reflect sounds to other workers and thus increase their exposure to detrimental noise levels.

12.5 REVERBERATION AND ABSORPTION

A review of Chapter 4 would be in order at this point to review the principles of reverberation time and the part that absorption plays in optimizing the reverberation time.

As we have seen, the location of the absorptive material is not specified by the principles of reverberation. Rather, it is from the analysis of the shape of the space that the optimum location can be determined.

From an architectural viewpoint, there are certain cardinal steps which must be followed. They are

1. From the volume and the anticipated prime use, determine the optimum reverberation time at 500 Hz. See Figure 4.6.
2. Determine the optimum reverberation time for other frequencies. See Figure 4.7.
3. For each of the frequencies of interest, calculate, using the reverberation formula, the amount of required absorption.

4. Establish the finish for each surface, that is, wood, plaster, acoustic tile, and so on.
5. From the table of absorption coefficients, determine the value for each surface.
6. Determine the total absorption by multiplying each area by its absorption coefficient and add all quantities.
7. Compare the value obtained in step (6) with that obtained in step (3).
8. Make adjustments in step (6) until the results desired in step (3) are obtained with a tolerance of 0.1 s.

Remember that no space can be optimized for more than one use. If more than one use, that is, a multipurpose auditorium, is contemplated, then a compromise reverberation goal should be sought. The compromise usually is an average of the optimum reverberation times for the various uses.

The latter principle is most important. A hall or classroom designed for speech is too dead for most music performances. Similarly, a hall designed for music is too reverberant for speech. Speech under these conditions loses its clarity and intelligibility. A major error occurring too often is the use of public address systems to improve speech in a space that is already too reverberant for voice. The introduction of a PA system merely garbles the voice more. PA systems are very useful when the space is too quiet without enough reverberation or when the projected distance to the audience for the performer is over 60 to 70 ft (20–25 m).

Application of the reverberation theory to commercial or industrial uses is possible but the economics of reducing noise levels in large spaces by adjusting the reverberation time are not very favorable. The amount of sound reduction achieved is usually not enough to provide relief from high noise levels. However, in restaurants the application of the principles of reverberation in a proper manner can be very effective in providing a pleasant environment.

12.6 MASKING

The principle of masking involves adding sound to an existing sound situation such that the new sound becomes predominant and thus the overriding audible sound. The masking of objectionable sounds with more acceptable sounds is commonly encountered in restaurants or other public gathering places. Unfortunately, the principle is abused when it is used to cover up poor architectural design.

There is however, a very positive use for the principle of masking. Several types of sound sources which are rated as objectionable noise do not lend themselves to either an engineering solution or to a solution which is practical and economic. To cite an example, many industries use long conveyor belts to move their product from one station to another. Bottling or canning industries are examples. In these situations the bottles or cans jostle each other as they are conveyed along the belt. This produces a continuous noise from one end of the line to another. Employees may have several stations along the line where they must be in contact with the product, either for inspection or for capping, and so on. Putting the conveyor line in an enclosure would solve the problem, but access would then be limited. Also the cost of enclosing the line is usually prohibitive. A solution successfully used in these cases is to supply music of a type acceptable to the employees at a sound level above the product noise (about 3 dB over the line level). The masking effect significantly reduces employee complaints.

In large offices designed around the open-space concept, a common source of complaint arises from multiple typewriter stations or computer consoles. Masking with music has again proved to be a satisfactory and economical solution.

Masking, while raising the noise level of existing sounds, has proven to be a very effective design tool when all other efforts at noise control have been abandoned for various reasons.

12.7 MODELING

A technique for analyzing the distribution of sound in large enclosed spaces such as auditoriums, concert halls, or large enclosed civic centers is the use of a scale model of the proposed structure. While the construction of a scale model is often a very expensive process, it can be justified when the acoustics of the enclosed space must meet a rigid criterion. The technique of using the model and gaining expertise in the analysis of the behavior of the sound wave is a rather sophisticated process and should only be used by someone with a considerable background in acoustics.

Two techniques are in common use; one uses pinpoint light sources, and the other a scaled up set of frequencies. The optical system places the pinpoint light sources in the model at the same location as the sound sources are expected to be in the full scale structure. By examining the reflections or diffractions of the light beams, the sound distribution can be visualized. The orientation of wall surfaces in the model can be altered to provide the most advantageous final shape.

The second technique involves using miniature sound speakers which propagate high frequency sound. The acoustic concept involved is contained in the following example. A scale model is made of a proposed music hall to be built using a scale of 1 ft to 10 ft (that is, 1:10 scale). This would require a shift upward of the comparable frequencies of 10:1. Thus the response of a 100-Hz tone in the actual building would be comparable to a 1000-Hz tone in the model. The response to 1000 Hz would be found at 10,000 Hz.

The difficulty with either modeling technique is to find materials, say for absorption, which behave the same at the high frequencies of the model as they do in the full-scale building. Therefore, the modeling technique is most valuable in those situations where various shape configurations are under study.

PROBLEMS

12.1. By use of a scale drawing, construct a spherical surface section and indicate, using ray diagrams, how sound is reflected. Using the same drawing, alter the spherical surface to one consisting of a series of interconnected staggered flat surfaces designed to diffuse the sound.

12.2. Assume that a classroom is to be analyzed for uniformity in sound distribution. The classroom measures 30 ft by 20 ft by 10 ft high. All surfaces are hard. Using a scale drawing of the plan and elevation of this room, show how the sound is propagated from:

a. A source in one corner at a height of 5 ft above the floor, and

b. The same source located in the center of the 20-ft wall at a height of 5 ft.

Where should absorbing material be placed to be most effective?

12.3. The wall of an auditorium has a series of bricks projecting out from the surface. These bricks, which measure 2 in. by 3 in. in cross-section, project out from the wall a distance of 1 in. at locations 1-ft apart. The architect claims that these bricks are deflectors of sound and thus tend to dispense the sound.

a. From your knowledge of the text material, is the architect's claim justified?

b. What frequencies will be affected by the projections?

12.4. Plans of a church require an acoustic analysis for reverberation time. The church floor plan measures 80 ft by 40 ft. The sidewalls are 20-ft high and the ridge line is 40 ft above the floor. The roof is triangular is shape. The church services require an unreinforced voice along with music as part of the services. The interim surfaces are to be specified by the acoustic consultant. A normal congregation consists of 250 people. Perform all of the functions necessary to provide the congregation with the most desirable acoustic environment.

APPENDIX A

Table of Abbreviations and Symbols

A	total absorption
A_c	acceleration
A_e	excess attenuation
A_f	total attenuation at a specific frequency
A_t	attenuation in dB/ft
A_{t2}	total attenuation in Room 2
A_1,A_2	pressure amplitudes
B_1,B_2	pressure amplitudes
B_n	number of blades
B_s	number of struts
B_v	blade tip velocity
C	velocity of sound (also c)
CNEL	Community Noise Equivalent Level
CNR	Composite Noise Rating
D	diameter
dB	decibel
dBA	decibel as measured on A scale
E	energy density
EPNL	Effective Perceived Noise Level
E_1	energy density in Room 1
E_2	energy density in Room 2

f	frequency
f_c	critical frequency
f_1	low frequency cutoff
f_m	midfrequency
f_n	lower band-pass frequency
f_{n+1}	upper band-pass frequency
f_N	frequency of the *n*th harmonic
f_p	frequency of peak attenuation
f_r	resonant frequency
f_s	frequency due to struts
°F	degrees Fahrenheit
F_d	number of aircraft flights, daytime
F_n	number of aircraft flights, nighttime
h	height of barrier above line of sight
H	total height of barrier
hp	horsepower
$i°$	incident angle
I	intensity
I_o	reference intensity
K	Kelvin
k	constant
$k_{1,2,...}$	wave number for wavelengths (1,2, . . .)
K	correction factor
$L^{1,2,...}$	sound pressure level in rooms (1,2, . . .) (dB)
L_{eq}	equivalent energy level (dB)
L_{dn}	day-night average level (dB)
l	wall thickness
l_m	muffler length
l_n	neck length
LF	loss factor of wall
LL	loudness level
m	meter
m_d	ratio of densities
m_s	ratio of muffler cross-section to duct cross-section
M_w	molecular weight
n	whole number
N	loudness level in NOYS
N_m	maximum level in NOYS
N_t	total level in NOYS
NEF	Noise Exposure Forecast
NNI	Noise and Number Index

NR	Noise Reduction
NRC	Noise Reduction Coefficient
P_A	mean square A-weighted sound pressure
P_e	perimeter
Ph	loudness level in phons
P_o	porosity
PNL	Perceived Noise Level (dB)
PNdB	perceived noise decibels
PWL	sound power level (dB)
P_a	atmospheric pressure
p_d	dynamic pressure
p_i	incident pressure
p_o	reference sound pressure
p_t	transmitted pressure
p_r	reflected pressure
p_{ra}	pressure at radial distance (x_r)
$p_{1,2,...}$	pressure in rooms (1,2, . . .)
RA	return air
R_{DC}	DC acoustic resistance
R_s	rotational speed (rpm)
r_a	specific acoustic resistance
r_p	ratio of maximum to minimum pressure
$r°$	reflected angle in degrees
S	area of radiating surface
S_a	cross-sectional area
S_o	reference surface area
S_w	area of intervening wall
So	loudness in sones
So_m	maximum loudness in sones
So_t	total loudness in sones
SPL	sound pressure level (dB)
$T°$	temperature
T_p	period
T_r	reverberation time
T_{rc}	corrected reverberation time below 500 Hz
T_{r500}	reverberation time at 500 Hz
TL	transmission loss
t_c	transmission coefficient
t_s	sample thickness
u_F	Fresnel function
u_i	incident particle velocity

u_r reflected particle velocity
u_t transmitted particle velocity

V total volume
V_m volume of material
V_o original volume
v velocity of radiating surface

W net surface density ($lb/ft^2/in.$)
W_o reference sound power
$W_{1,2,\ldots}$ sound power in room (1,2, . . .)

x displacement (distance)
x_a distance, source to barrier
x_A distance, source to top of barrier
x_{ab} straight line distance, source to receiver
x_{AB} diffracted distance, source to receiver
x_{ar} specific acoustic reactance
x_b distance, barrier to receiver
x_B distance, top of barrier to receiver
x_D mean free path
x_F Fresnel integral
x_m maximum displacement (amplitude)
x_r radial distance (radius)
x_{rs} simulated radius

y_F Fresnel integral

$z_{1,2,\ldots}$ acoustic impedance in medium (1,2, . . .)

α absorption coefficient
α_A absorption after installation
α_B absorption before installation
α_n absorption coefficient at normal incidence
α_r absorption coefficient at random incidence
α_R absorption coefficient for rooms
α_T absorption coefficient for an impedance tube
γ ratio of specific heats
e density
ρc specific acoustic impedance
λ wavelength

APPENDIX B

Procedure for Adding Several Decibel Levels

1. Convert the decibel levels to be added to *power ratios*. See the dB to Power Ratio Conversion Table on page 184.
2. Add the power ratios corresponding to the decibel levels to be added. For example

Add 87 dB + 89 dB + 100 dB

dB	Power Ratio
87	0.5012
89	0.7943
100	10.0000
	11.2955

3. Locate the decibel level corresponding to the total power ratio. In the example above: The total power ratio is equal to 11.2955. The corresponding decibel level for 11.2955 is between 100.5 and 101.0 dB. Choose the closest dB value which corresponds to 11.2955; in this example, 100.5 dB. *Note:* The dB to Power Ratio Conversion Table covers the range between 70 and 119.5 dB. Other lower and higher values may also be obtained. For

each 10-dB increment the power ratio changes by a factor of 10. For example, a 60-dB power ratio is 0.0010, a 120-dB power ratio is 1000, and so on.

dB To Power Ratio Conversion Table

dB	Power Ratio	dB	Power Ratio	dB	Power Ratio
70.0	0.01000	87.0	0.5012	104.0	25.12
70.5	0.01122	87.5	0.5623	104.5	28.18
71.0	0.01259	88.0	0.6310	105.0	31.62
71.5	0.01413	88.5	0.7079	105.5	35.48
72.0	0.01585	89.0	0.7943	106.0	39.81
72.5	0.01778	89.5	0.8913	106.5	44.67
73.0	0.01995	90.0	1.000	107.0	50.12
73.5	0.02239	90.5	1.122	107.5	56.23
74.0	0.02512	91.0	1.259	108.0	63.10
74.5	0.02818	91.5	1.413	108.5	70.79
75.0	0.03162	92.0	1.585	109.0	79.43
75.5	0.03548	92.5	1.778	109.5	89.13
76.0	0.03981	93.0	1.995	110.0	100.0
76.5	0.04467	93.5	2.239	110.5	112.2
77.0	0.05012	94.0	2.512	111.0	125.9
77.5	0.05623	94.5	2.818	111.5	141.3
78.0	0.06310	95.0	3.162	112.0	158.5
78.5	0.07079	95.5	3.548	112.5	177.8
79.0	0.07943	96.0	3.981	113.0	199.5
79.5	0.08913	96.5	4.467	113.5	223.9
80.0	0.1000	97.0	5.012	114.0	251.2
80.5	0.1122	97.5	5.623	114.5	281.8
81.0	0.1259	98.0	6.310	115.0	316.2
81.5	0.1413	98.5	7.079	115.5	354.8
82.0	0.1585	99.0	7.943	116.0	398.1
82.5	0.1778	99.5	8.913	116.5	446.7
83.0	0.1995	100.0	10.00	117.0	501.2
83.5	0.2239	100.5	11.22	117.5	562.3
84.0	0.2512	101.0	12.59	118.0	631.0
84.5	0.2818	101.5	14.13	118.5	707.9
85.0	0.3162	102.0	15.85	119.0	794.3
85.5	0.3548	102.5	17.78	119.5	891.3
86.0	0.3981	103.0	19.95		
86.5	0.4467	103.5	22.39		

APPENDIX C

Fresnel Integrals

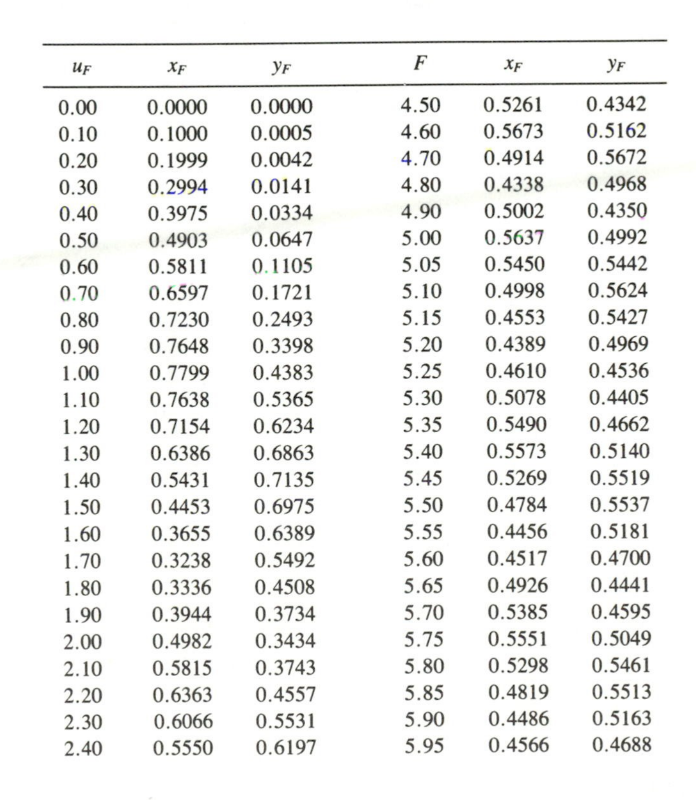

u_F	x_F	y_F	F	x_F	y_F
0.00	0.0000	0.0000	4.50	0.5261	0.4342
0.10	0.1000	0.0005	4.60	0.5673	0.5162
0.20	0.1999	0.0042	4.70	0.4914	0.5672
0.30	0.2994	0.0141	4.80	0.4338	0.4968
0.40	0.3975	0.0334	4.90	0.5002	0.4350
0.50	0.4903	0.0647	5.00	0.5637	0.4992
0.60	0.5811	0.1105	5.05	0.5450	0.5442
0.70	0.6597	0.1721	5.10	0.4998	0.5624
0.80	0.7230	0.2493	5.15	0.4553	0.5427
0.90	0.7648	0.3398	5.20	0.4389	0.4969
1.00	0.7799	0.4383	5.25	0.4610	0.4536
1.10	0.7638	0.5365	5.30	0.5078	0.4405
1.20	0.7154	0.6234	5.35	0.5490	0.4662
1.30	0.6386	0.6863	5.40	0.5573	0.5140
1.40	0.5431	0.7135	5.45	0.5269	0.5519
1.50	0.4453	0.6975	5.50	0.4784	0.5537
1.60	0.3655	0.6389	5.55	0.4456	0.5181
1.70	0.3238	0.5492	5.60	0.4517	0.4700
1.80	0.3336	0.4508	5.65	0.4926	0.4441
1.90	0.3944	0.3734	5.70	0.5385	0.4595
2.00	0.4982	0.3434	5.75	0.5551	0.5049
2.10	0.5815	0.3743	5.80	0.5298	0.5461
2.20	0.6363	0.4557	5.85	0.4819	0.5513
2.30	0.6066	0.5531	5.90	0.4486	0.5163
2.40	0.5550	0.6197	5.95	0.4566	0.4688

(continued)

u_F	x_F	y_F	F	x_F	y_F
2.50	0.4574	0.6192	6.00	0.4995	0.4470
2.60	0.3890	0.5500	6.05	0.5424	0.4689
2.70	0.3925	0.4529	6.10	0.5495	0.5165
2.80	0.4675	0.3915	6.15	0.5146	0.5496
2.90	0.5626	0.4101	6.20	0.4676	0.5398
3.00	0.6058	0.4963	6.25	0.4493	0.4954
3.10	0.5616	0.5818	6.30	0.4670	0.4555
3.20	0.4664	0.5933	6.35	0.5240	0.4560
3.30	0.4058	0.5192	6.40	0.5496	0.4965
3.40	0.4385	0.4296	6.45	0.5292	0.5398
3.50	0.5326	0.4152	6.50	0.4816	0.5454
3.60	0.5880	0.4923	6.55	0.4520	0.5078
3.70	0.5420	0.5750	6.60	0.4690	0.4631
3.80	0.4481	0.5656	6.65	0.5161	0.4549
3.90	0.4223	0.4752	6.70	0.5467	0.4915
4.00	0.4984	0.4204	6.75	0.5302	0.5362
4.10	0.5738	0.4758	6.80	0.4831	0.5436
4.20	0.5418	0.5633	6.85	0.4539	0.5060
4.30	0.4494	0.5540	6.90	0.4732	0.4624
4.40	0.4383	0.4622	6.95	0.5207	0.4591

APPENDIX D

Derivation of the Modified Fehr Equation (3–6)

Start with Equation (3–7)

$$\text{SLR} = 10 \log (10N)$$

where

$$N = \frac{2}{\lambda}\left[R\left(\sqrt{1 + \frac{H^2}{R^2}} - 1\right) + D\left(\sqrt{1 + \frac{H^2}{R^2}} - 1\right)\right]$$

for $H < R$ and $H < D$, use the first two terms of a Taylor-series expansion

$$\sqrt{1 + X} = 1 + \frac{1}{2}X - \frac{1}{8}X^2 + \frac{1}{16}X^2 + \cdots$$

then

$$N \simeq \frac{2}{\lambda}\left[R\left(1 + \frac{1}{2}\frac{H^2}{R^2} - 1\right) + D\left(1 + \frac{1}{2}\frac{H^2}{D^2} - 1\right)\right]$$

$$= \frac{2}{\lambda}\left[\frac{1}{2}\frac{H^2}{R} + \frac{1}{2}\frac{H^2}{\text{D}}\right]$$

$$= \frac{H^2}{\lambda R} + \frac{H^2}{\lambda D} = \frac{H^2}{\lambda}\left[\frac{1}{R} + \frac{1}{D}\right]$$

Now, if $R \ll D$ then the first term predominates an

$$N \simeq \frac{H^2}{\lambda R}$$

so the modified Fehr equation becomes

$$\mathrm{SLR} = 10 \log \left(\frac{10H^2}{\lambda R} \right)$$

where $R \ll D$ and $H < R$.

Index

D

E

F

G

H

I

L

M

N

O

P

R

S

WIDENER UNIVERSITY
WOLFGRAM
LIBRARY
CHESTER, PA.